Praise for *Hands-On APIs for AI and Data Science*

Hands-On APIs for AI and Data Science is an awesome contribution to the data science community. Day provides a structured guide to a core topic data scientists often learn too late: delivering solutions to users. You'll master APIs, but along the way, you'll also add a dozen more tools to your data science toolbox.

Do yourself a favor—add this book to your data science library for your continued professional development and become a better, more effective data scientist!

—*Alex Gutman, author of* Becoming a Data Head: How to Think, Speak, and Understand Data Science, Statistics, and Machine Learning

Day takes the reader through a thorough, yet clear, roadmap of building APIs, using an extremely topical industry (fantasy football) as the example. This is essential reading for anyone looking to round out their data science capabilities as an individual or better serve their customers as a company.

—*Eric Eager, vice president of football analytics,* Carolina Panthers

With the growing importance of APIs in data science and AI, this book is an essential resource for gaining practical insights. It prioritizes understanding your consumers, which is essential for designing and building great APIs. This book is filled with actionable examples and expert guidance. It is an invaluable read for anyone looking to create impactful APIs.

—*James Gough, author of* Mastering API Architecture

This book is a fantastic resource for data scientists who need to use APIs, whether you're building them or accessing data through them. It's very practical, it's fun to read, and it'll be extremely useful to any data scientist who wants to improve their software engineering skills.

—*Catherine Nelson, author of*
Software Engineering for Data Scientists

Ryan offers a comprehensive guide for data scientists at all levels that blends deep technical expertise with practical strategies on mastering API usage.

—*Kris Rowley, CSBS chief data officer and*
Data Foundation board member

Hands-On APIs for AI and Data Science avoids the biggest mistake I see in technical books: it provides practical lessons grounded in how technology is actually used. With fun examples from sports data, author Ryan Day walks through multiple angles of a single API project. Anyone in data science would be wise to build their career on the foundations Ryan has laid out in this book.

—*Adam DuVander, EveryDeveloper*

Ryan does a great job teaching you how to both be a better API user and creator using engaging examples from fantasy sports.

—*Richard Erickson, data scientist and O'Reilly author of*
Football Analytics with Python and R

Ryan Day's book is an essential resource for football analytics professionals, from newcomers to seasoned experts. This book equips readers with the tools to build and deploy APIs that power advanced data workflows, from player performance modeling to real-time fantasy football applications. With Ryan's guidance, you'll learn to integrate APIs into your analytics toolbox and take your insights to the next level.

—*Amelia Probst, data scientist, Pro Football Focus*

If you're looking to skill up on APIs and understand how important they are to building effective AI applications, this book delivers a mix of theory and hands-on exercise to get you there.

—*Jeff Frederickson, software engineer*

This book is a must-have for Python developers seeking to build powerful and efficient APIs, utilizing the latest FastAPI technology. With clear explanations and practical examples, it guides readers through every step of API creation and deployment, making complex topics approachable and actionable.

—*Megan Silvey, founder and data science consultant,*
Silvey Solutions

Hands-On APIs for AI and Data Science is an essential read for today's data and IT professionals aiming to keep pace in our ever-evolving data-driven world. Ryan's ability to present complex concepts through hands-on application makes it easy for readers to apply what they've learned in practice, in real-world scenarios, or even on the job. Highly recommended for beginners and seasoned professionals, and you may even learn a little fantasy football along the way!

—*Richard Bright, enterprise data architect*

Hands-On APIs for AI and Data Science

Python Development with FastAPI

Ryan Day

O'REILLY®

Hands-On APIs for AI and Data Science

by Ryan Day

Published by O'Reilly Media, Inc., 1005 Gravenstein Highway North, Sebastopol, CA 95472.

O'Reilly books may be purchased for educational, business, or sales promotional use. Online editions are also available for most titles (*http://oreilly.com*). For more information, contact our corporate/institutional sales department: 800-998-9938 or *corporate@oreilly.com*.

Acquisitions Editor: Michelle Smith	**Indexer:** Sue Klefstad
Development Editor: Corbin Collins	**Interior Designer:** David Futato
Production Editor: Aleeya Rahman	**Cover Designer:** Karen Montgomery
Copyeditor: Tove Innis	**Illustrator:** Kate Dullea
Proofreader: Audrey Doyle	

March 2025:   First Edition

Revision History for the First Edition

2025-03-04:   First Release

See *http://oreilly.com/catalog/errata.csp?isbn=9781098164416* for release details.

978-1-098-16441-6

[LSI]

For Allison

Table of Contents

Part II.   Using APIs in Your Data Science Project

Part III. Using APIs with Artificial Intelligence

Preface

To succeed in AI, first master APIs. Becoming skilled at APIs is more valuable than ever, thanks largely to the growth of artificial intelligence, machine learning, and data science.

But learning a such wide-ranging skill is intimidating. How is it to be done? You can take comfort in the fact that you don't have to learn every skill, and certainly not all at once. Pick up one skill at a time through hands-on practice. Each skill you learn makes the next one easier, like building blocks.

Why Should You Read This Book?

If you're reading this book, you want to build your skills. I have found that the best way to do that is through hands-on coding. If you do your coding in the open by publishing your code in a public repository and blogging and sharing what you create, you can pass along your knowledge to help others. You'll also build a solid portfolio of work that provides a concrete demonstration of your expertise to employers.

Who This Book Is For

Since this book sits at the intersection of APIs, AI, and data science, it will be valuable to several types of readers.

Data Scientists

Data scientists use APIs all the time, so there's a temptation to think there's nothing new to learn about using APIs. Isn't calling an API just a few lines of code? It's true that making one call to a REST API is a simple task, which is certainly a reason they have become so prevalent. But using an API in a way that is robust and resilient—and that doesn't cause problems for the provider—requires more care.

This book will teach some techniques you may not have learned yet, such as:

- Developing and deploying APIs
- Creating software development kits (SDKs) and API clients
- Creating and publishing Python packages to PyPI
- Publishing machine learning models as APIs
- Creating Streamlit data apps
- Creating Airflow data pipelines
- Creating generative AI applications using LangChain and ChatGPT

API Developers and Designers

API developers and designers can learn how to enhance their APIs for important new audiences. They'll learn about data scientists: the jobs they do, the tasks they perform, and the API features they love. They'll also learn about generative AI applications: how they call APIs and what features they need in an API.

And the hands-on examples will teach a variety of new skills:

- Creating Python APIs with FastAPI, SQLAlchemy, and Pydantic
- Containerizing APIs with Docker
- Deploying APIs to cloud hosts
- Creating Python SDKs and publishing to PyPI
- Creating generative AI applications using LangChain and ChatGPT

Job Seekers and Role Changers

The skills above are valuable in the marketplace, so learning them can help you find your first role or a new role in data science or software development. This book is arranged around building portfolio projects, which will give you specific goals to accomplish, and tangible evidence of your work.

Creating Portfolio Projects

While completing the book, you'll create three portfolio projects that you can publish in GitHub repositories to show the work you've done. Table P-1 explains the purpose and source repository you will use as a basis for your project.

Table P-1. New tools or services used in this chapter

Project	Purpose
Part I portfolio project	Creating a Python API and SDK
Part II portfolio project	Creating data science apps using Streamlit, Airflow, and Jupyter Notebooks
Part III portfolio project	Creating a machine learning API and generative AI application using scikit-learn, ONNX, LangChain, and ChatGPT

> When you complete your projects, please reach out and let me know at *ryan@handsonapibook.com* or on LinkedIn (*https://www.linkedin.com/in/ryanday1*) so I can celebrate your accomplishment with you. I look forward to seeing what you build.

Using This Book

Instead of reading this book from beginning to end, I recommend that you pick the skill you want to learn and start working through the chapter that teaches it. You can do this quickly in the following way:

1. Decide whether you want to start by creating an API (Part I), using APIs with data science (Part II), or using APIs with AI (Part III).

2. Follow the instructions in the introductory chapter of that part to clone the GitHub repository and launch a Codespace.

3. Follow the instructions in the relevant chapter and run the code.

If I've done my job properly, each chapter can stand on its own. After you've learned one skill, look around and find another skill you want to learn, and do the same. The skills in this book are like building blocks: each piece you learn prepares you to understand the other parts more deeply. All of them together give you quite a substantial understanding of APIs in data science and AI.

What This Book Is Not

This book doesn't teach Python syntax to beginners. You will get the most from the coding examples if you have a foundational knowledge of Python. Although you can probably get the code to work by following the steps in each chapter, I suggest you begin with one of the excellent introductory Python books that O'Reilly publishes, such as *Introducing Python, 3rd Edition*, by Bill Lubanovic.

This book also assumes a basic understanding of using the command-line terminal in Linux or Unix. You don't need to be a Linux administrator, but you should be familiar with running commands in the terminal as a developer. (All the steps are explained, but when you run into an issue, you might get frustrated without some background in that environment.)

The book introduces several useful Python frameworks such as FastAPI, Pydantic, Streamlit, Airflow, and LangChain. However, it does not address detailed topics necessary to run them in a *production* environment, such as security, performance, and infrastructure. I hope that you'll enjoy the projects in this book enough that you'll continue your learning using the references that I mention at the end of each chapter.

Keep in mind that the services and tools in this book are changing rapidly, so depending on when you are reading this, some of the steps and figures may look a bit different.

Why Fantasy Football?

> If you were to sit down and rank hobbies that people obsess over the details of, fantasy sports and software development would both be near the top of that list. When you combine these hobbies, the possibilities are endless!
>
> —Joey Greco, creator of the Leeger stats application

The portfolio projects in the book follow the story of an imaginary fantasy sports league host website: SportsWorldCentral.com. Through your projects, you will design and build APIs for data-focused users, then switch roles and build data science and generative AI applications using the APIs you created.

As Joey Greco says so eloquently, fantasy sports was the natural choice for the scenario. (You'll hear more from him later in the book.) Fantasy sports is a natural playground for data scientists, and fantasy websites use plenty of APIs. I've spent many hours over the years wading into both of those as a devoted (or addicted) fantasy manager. Fantasy managers are also fast adopters of any predictive or prescriptive analytics feature the fantasy websites give them. (If you doubt it, you haven't watched a manager pick up two free agents and make three lineup changes to push their win probability from 45% to 53%.)

Fantasy sports are a fun way to geek out on the overlap between AI, data science, and APIs. As you code your way through the book, I hope you have as much fun as I did writing it.

Get More Tips on APIs, AI, and Data Science

The content in this book will be a really solid foundation for these topics, and I hope it raises your interest in learning more. To get more tips in this subject, you can subscribe to my newsletter by visiting *https://handsonapibook.com*.

Conventions Used in This Book

The following typographical conventions are used in this book:

Italic
> Indicates new terms, URLs, email addresses, filenames, and file extensions.

`Constant width`
> Used for program listings, as well as within paragraphs to refer to program elements such as variable or function names, databases, data types, environment variables, statements, and keywords.

`Constant width bold`
> Shows commands or other text that should be typed literally by the user.

`Constant width italic`
> Shows text that should be replaced with user-supplied values or by values determined by context.

> This element signifies a tip or suggestion.

> This element signifies a general note.

> This element indicates a warning or caution.

Using Code Examples

Supplemental material (code examples, exercises, etc.) is available for download in three repositories, one for each part:

- *Part I: https://github.com/handsonapibook/api-book-part-one*
- *Part II: https://github.com/handsonapibook/api-book-part-two*
- *Part III: https://github.com/handsonapibook/api-book-part-three*

If you have a technical question or a problem using the code examples, please send email to *support@oreilly.com*.

This book is here to help you get your job done. If example code is offered with this book, you may use it in your programs and documentation. You do not need to contact us for permission unless you're reproducing a significant portion of the code. For example, writing a program that uses several chunks of code from this book does not require permission. Selling or distributing examples from O'Reilly books does require permission. Answering a question by citing this book and quoting example code does not require permission. Incorporating a significant amount of example code from this book into your product's documentation does require permission.

We appreciate, but generally do not require, attribution. An attribution usually includes the title, author, publisher, and ISBN. For example: "*Hands-On APIs for AI and Data Science* by Ryan Day (O'Reilly). Copyright 2025 Ryan Day, 978-1-098-16441-6."

If you feel your use of code examples falls outside fair use or the permission given above, feel free to contact us at *permissions@oreilly.com*.

O'Reilly Online Learning

For more than 40 years, *O'Reilly Media* has provided technology and business training, knowledge, and insight to help companies succeed.

Our unique network of experts and innovators share their knowledge and expertise through books, articles, and our online learning platform. O'Reilly's online learning platform gives you on-demand access to live training courses, in-depth learning paths, interactive coding environments, and a vast collection of text and video from O'Reilly and 200+ other publishers. For more information, visit *https://oreilly.com*.

How to Contact Us

Please address comments and questions concerning this book to the publisher:

O'Reilly Media, Inc.
1005 Gravenstein Highway North
Sebastopol, CA 95472
800-889-8969 (in the United States or Canada)
707-827-7019 (international or local)
707-829-0104 (fax)
support@oreilly.com
https://oreilly.com/about/contact.html

We have a web page for this book, where we list errata, examples, and any additional information. You can access this page at *https://oreil.ly/hands-on-api*.

For news and information about our books and courses, visit *https://oreilly.com*.

Find us on LinkedIn: *https://linkedin.com/company/oreilly-media*

Watch us on YouTube: *https://youtube.com/oreillymedia*

Acknowledgments

I'm sure I'll miss a few people, but thank you so much to everyone who has interacted with this content in the past year in articles, newsletters, and presentations. Your feedback has made the content better, and your encouragement and interest have kept me going. Extra-special thanks to the Data Science KC and Data Community DC for all of your feedback.

I'm so grateful to the O'Reilly team who guided me through the process of writing this book. Thank you so much to the entire editorial staff. Special thanks to Michelle Smith, the acquisitions editor, and Corbin Collins, the development editor, for believing in the book and getting me to the finish line. Thank you, Chris Faucher, Aleeya Rahman, and Tove Innis, for your professional handling of the authoring process.

I had so much fun interviewing experts for this book, and many of their stories are included in the chapters. Thanks to Joey Greco, Kyle Borgognoni, Zan Markhan, Alexandre Airvault, Francisco Goitia, Simon Yu, Robin Linacre, Bill Doerrfield, Samuel Colvin, Frank Kilcommins, Kade Halabuza, and Jim Higginbotham for sharing your expertise and enthusiasm for the technology and data. Thanks to Keith McCormick, who showed me the opportunities available in data science.

Thank you so much to my technical reviewers: Richard Bright, Richard Erickson, Jeff Fredrickson, Amelia Probst, and Megan Silvey. You saved me from many embarrassing mistakes, and your technical expertise helped make the book much more accurate and valuable. Extra thanks to Richard Erickson, who guided me through the book proposal process and publishing process, in addition to being an active technical reviewer.

Thanks to Kris Rowley, Ngoc Vu, and all the staff at CSBS who were excited for me and encouraged me to write the book.

Most of all, I thank God for the blessings and opportunities given to me, beginning with a mom and dad who taught me to think clearly. Thank you to my wife, Allison, who was my constant encourager from beginning to end. Thank you to all my family for their love and support. Ethan, Myles, Sam, Cara, Gabby, and Gabe, I love you all.

Building APIs for Data Science

Part I of this book walks you step by step through example projects to build APIs using Python and FastAPI and deploy them to the cloud:

- In Chapter 1, you will begin your portfolio project by understanding the user needs and selecting use cases to fulfill.

- In Chapter 2, you will select an API architecture and start creating an API to fulfill the needs of data scientists.

- In Chapter 3, you will create your database using SQLite, create Python code to read the database with SQLAlchemy, and perform unit testing with Pytest.

- In Chapter 4, you will create the FastAPI code to use this data and publish it as a REST API.

- In Chapter 5, you will document your API using FastAPI's built-in capabilities.

- In Chapter 6, you will deploy your API to the cloud using Render and Amazon Web Services (AWS).

- In Chapter 7, you will create a software development kit (SDK) to make your API easier for Python developers to use.

## CHAPTER 1

Creating APIs That Data Scientists Will Love

Application programming interfaces (APIs) are important to data scientists. But how often are data scientists considered by API designers and developers? Data scientists frequently use APIs as data sources for their work. They have some needs that are different than those of software developers or other consumers. If API producers want to make data scientists happy, they will do well to serve these needs.

How Do Data Scientists Use APIs?

The Anaconda State of Data Science Report (*https://oreil.ly/anaconda*) found that data scientists spend the bulk of their time performing three main activities: preparing or cleansing data (38%); creating reports, presentations, or data visualizations (29%); and selecting, training, and deploying models (27%). This book demonstrates how data scientists use APIs for these tasks.

Preparing or cleansing data often occurs when data scientists perform *exploratory data analysis (EDA)* on a new dataset by analyzing its contents, formats, and patterns. Other times, this work is part of a scheduled *data pipeline*, which is a sequence of software tasks that pull from multiple data sources and reformat or remove errors from the data so that it can be used downstream in visualizations, reports, or models. *Data engineer* is a another job title for people who specialize in these tasks. Chapter 9 will demonstrate using APIs in exploratory data analysis. Chapter 10 will demonstrate using APIs in data pipelines.

Creating reports and visualizations are important activities to demonstrate insights gleaned from data and present the data in a way that enables an organization to monitor its operations or make better decisions. These are also referred to as *analytics*. Data scientists make calls to APIs to provide data for a variety of analytics products. *Data analyst* is another job title for people who specialize in these tasks. Chapter 9 will show how data scientists create visualizations using Python in Jupyter Notebooks, and Chapter 11 will show how interactive visualizations can be deployed in applications using Python and Streamlit.

Training and deploying data science models are activities that use machine learning or other mathematical techniques to make predictions, cluster data into groups, perform natural language processing, and accomplish a variety of other tasks. The use of APIs here can be divided into two buckets: API consumers and API producers. As consumers, data scientists use APIs as input sources to their models and call machine learning APIs to perform tasks. As producers, they deploy their own models as APIs for others to use. *Machine learning engineer* is another job title for people who specialize in deploying models, and Chapter 13 will demonstrate how to deploy a machine learning model using FastAPI and Python.

> Application programming interface (API) is a term that has a variety of meanings in computer programming. This book will focus on web APIs, which are software programs that you call using HTTP to retrieve data or execute commands. API producers develop and host an API for internal consumers inside their network or for external consumers on the internet. API consumers send a request to an API. The API receives this request and sends back a response. This is API communication.

What Tools Do Data Scientists Use?

Data scientists use a range of commercial and open source software tools to perform their work, along with commercial cloud platforms. According to the State of Data Science report, the most common programming languages that data scientists use are Python (58% always or frequently use it) and SQL (42% always or frequently use it). The R language did not rank highly in this survey, but in my experience it is used by many data scientists. R is popular in football analytics thanks to the nflverse (*https://github.com/nflverse*) collection of data and R packages. The program code in this book will use both Python and SQL.

For a development environment, data scientists typically are comfortable using command-line tools (such as the Bash shell for Linux or PowerShell for Windows). They use IDEs such as Visual Studio Code (VS Code), PyCharm, or RStudio. Data scientists also use notebook environments, which are unique interactive programming environments that allow Markdown descriptions to be interlaced with program code and output of commands.

Data scientists use a variety of tools to maintain standalone development environments, such as the venv and conda libraries. Docker in an important tool for packaging environments and deploying applications. *Dev containers* build on Docker to

---

provide full-featured virtual environments with an embedded IDE—GitHub Codespaces is an example that you will use in this book.

Designing APIs for Data Scientists

Now that you know a little bit about the tasks data scientists perform and the tools they like to use, here are some tips for designing APIs they will love. This section will focus on Python, but many of the tips are also appropriate for the R language:

Your API should always return data in JSON format.
Your API endpoints should return data in JSON format, rather than XML. The Python ecosystem has strong support for JSON, which is a lightweight data format that supports hierarchies and is human-readable. JSON can be converted into lists and dictionaries, which are fundamental Python data structures. Most web APIs return JSON as a standard already, so this isn't much of a stretch.

You should provide an SDK to consume the API.
While Python can call APIs directly from within the code, data scientists are accustomed to installing Python libraries using commands such as `pip` and `conda`. Publishing a Python library will make life easier for the data scientists and allow you to enforce good coding practices in the way your API is called.

You should provide standard external identifiers in your data.
Data scientists often combine data from multiple sources in their visualizations and data pipelines. If you provide industry-standard identifiers, it will allow the data scientist to more easily join data sources.

Your data should conform to data type definitions.
Data scientists use API data for performing calculations and creating models, so data type validity is important. Where an invalid character in an API can probably display fine on a web page, an invalid number in a numeric field could make a record unusable for data science tasks. So, data returned from APIs must conform to its definition, such as an OpenAPI Specification (OAS) file.

You should provide a method for bulk downloads.
When data scientists are analyzing a new dataset or training a machine learning model, they often examine the full contents. Because a full dataset may be quite large, making an API call may strain the resources of the API as well as the data scientist's local environment. Timeouts and memory overflows can cause frustration and delays. If the API provider provides a bulk download capability, this streamlines this analysis. The bulk download is also useful when a data scientist is performing the initial full load of a new data pipeline. Useful data formats for data scientists include *comma-separated values* (CSV) and Apache Parquet (*https://oreil.ly/UYELV*).

Your API should support querying by last changed date.

After data scientists have performed the initial full load of a data pipeline, they want to process regular updates. A typical data pipeline schedule would be to run a recurring job daily to process the *deltas*, or records that have changed. If the API provides a last changed date query parameter, this allows the data pipeline to retrieve any new records or updated records.

Introducing Your Part I Portfolio Project

You will address all of these design tips as you build your Part I portfolio project, which demonstrates your ability to develop an API and an SDK. Many of these are automatically addressed for you by the tools you have chosen, and a few you will custom-code yourself.

Here is an overview of the work ahead of you:

- Chapter 1: Understanding your users and selecting the right API
- Chapter 2: Selecting your API architecture and setting up your development environment
- Chapter 3: Creating your database
- Chapter 4: Developing the FastAPI code
- Chapter 5: Documenting your API
- Chapter 6: Deploying your API to the cloud
- Chapter 7: Creating an SDK for your API

You will follow a realistic scenario in each chapter, with each step building on the previous steps. It's time to dive into that scenario now, from an industry that is close to my heart: fantasy sports.

Every API Has a Story

Every API has a story and a reason it gets built. The story of some APIs is straight out of the movie *Field of Dreams*: "If you build it, they will come." A company builds an API without performing any research on the potential users—sometimes customers come, and other times an API languishes unused and unremembered. In contrast, successful APIs fulfill the needs of real consumers. That's the case with the API you will build in this book.

Meeting Your Company: SportsWorldCentral

You are a software developer working for a website named SportsWorldCentral, SWC for short. SWC provides sports news and also hosts fantasy games such as fantasy football and fantasy soccer. SWC's customers are sports fans who join up in a league with friends and draft real-world players onto their teams. They watch real-world games or matches, and when the players on their teams do well, they score points for their fantasy team. SWC helps all those owners keep track of their teams and gives live scoring updates each week as fantasy managers win or lose.

SWC's fantasy football website contains information about an entire fantasy football league, such as shown in Figure 1-1.

| League Home | Stats | Matchups | Transactions | My Team | Support |

SportsWorldCentral Fantasy Football

Week 6 Matchup

	Proj. Pts	Win %		Proj. Pts	Win %
Long Lost Lizards	98.7	42%	Touchdown Teresas	104.2	58%

My Team
Long Lost Lizards

Record
3-1-1, 2nd

Next Game
Touchdown Teresas, 3-1-1

League Standings: Crazy Rotten Fantasy League

Rank	Team	W-L-T	Points For	Points Against
1	Lincoln Lawyers	4-1-0	527	480
2	Long Lost Lizards	3-1-1	550	522
3	Touchdown Teresas	3-1-1	570	533
4	Fumble Tumble	3-2	490	399
5	Watch Me Whip	3-2	524	455
6	Kick Seven	2-3	501	502
7	Farcical Footballers	2-3	481	492

Figure 1-1. SWC League home page

The website also has a large amount of detail about individual fantasy teams, as shown in Figure 1-2.

SportsWorldCentral Fantasy Football

My Team: Long Lost Lizards

Roster Spot	Position	Player	Team	Bye	Points Scored
QB	QB	Hurts, Jalen	Phi	4	101
RB	RB	Ekeler, Austin	LAR	3	40
RB	RB	Pacheco, Isaiah	KC	7	77
WR	WR	Brown, A.J.	Phi	4	80
WR	WR	Addison, Jordan	Min	8	37
Flex	WR	Flowers, Zay	Bal	4	42
TE	TE	LaPorta, Sam	Det	6	65
PK	PK	Butker, Harrison	KC	7	48
Bench	QB	Goff, Jared	Det	6	97
Bench	WR	Sutton, Cortland	Den	9	38
Bench	WR	Mims, Marvin	Den	9	37
Bench	WR	Williams, Jameson	Det	6	0
Bench	RB	Elliott, Ezekiel	Dal	10	44
Bench	RB	Edwards, Gus	Bal	4	77

Figure 1-2. SWC My Team page

SWC is an imaginary company, but fantasy sports is a very real and thriving entertainment industry. There are around 40 million fantasy football managers, according to fantasy industry estimates, making fantasy football the largest US fantasy sport. Fantasy soccer is popular worldwide. For example, more than 10 million active managers compete on the Fantasy Premier League website (*https://fantasy.premierlea gue.com*), which follows just one of the major international soccer leagues. Real-world examples of fantasy league host websites include Yahoo.com, Sleeper, MyFantasy League.com, and many others.

In addition to fantasy league web hosts, fantasy managers pay for subscriptions to fantasy advice websites such as The Fantasy Footballers (*https://oreil.ly/Hu6Pn*), PFF (*https://oreil.ly/qEVlz*), and FantasyPros (*https://oreil.ly/-9x4F*). These websites provide highly customized analytical products such as dashboards, charts, predictive models, and recommendation engines to help fantasy managers run their teams. To customize this content to a fantasy manager's team, the fantasy league host needs to provide an API that the fantasy advice websites can consume.

API Perspectives: Kyle Borgognoni on Fantasy Football APIs

Kyle Borgognoni is the editor-in-chief of The Fantasy Footballers fantasy advice website (*https://www.thefantasyfootballers.com*) and podcast, which has won multiple fantasy industry awards for popularity and accuracy. Kyle is also host of the *Fantasy Footballers Dynasty* podcast (*https://oreil.ly/FFDP*) and a writer for the website.

What type of advice content do you provide on your website?

We provide general fantasy statistics like fantasy points, targets, yards, and red zone numbers; advanced fantasy stats like routes, targets per route run, and efficiency metrics; and performance-based stats like consistency scores, career performance, and strength of schedule. We also provide the Ultimate Draft Kit, which has projections and stat lines for every player in the NFL; the Dynasty Pass that has college production profiles and metrics from the NFL combine; and the Daily Fantasy Pass.

How do you gather all of the data for the products you provide?

We get a lot of information from data brokers and partners like SportsData. We also use APIs from websites like Sleeper and Yahoo!. For the podcast, it is a combination of data scraping and endless worksheets that I have built.

How does it benefit a league host website to provide APIs for advice websites to use?

To use Sleeper as an example, they are providing an API because it helps their platform. With their data, folks can build useful tools and create an ecosystem around their service that provides additional value to their users for free. If users come to our platform, and they can easily import their Sleeper teams, see Sleeper ADP, or whatever, they are more likely to use Sleeper to host their fantasy leagues. This is why they provide their proprietary data for free in an easy-to-use API. This is valuable to us because our business is selling these tools to users.

SWC Needs an API

The web team that manages the SWC website has seen a rising amount of traffic on the Fantasy Football web pages from *web scrapers*, which are computer programs that read websites and extract the data from them. Instead of seeing this as a problem, the savvy SWC product owner realizes this is an indicator of a potential market: people want to use the SWC Fantasy Football data.

The product owner reaches out across the company to identify any other requests received for data access or data sharing. The help desk finds dozens of customer support tickets asking why many of the advice websites can't import their league data from SWC—indicating a need to share data with those sites. Additional customer tickets are from data scientists and other data-savvy users requesting direct access to the football data from the website so they can create custom dashboards and metrics.

Independent mobile app developers have also offered to develop mobile apps to serve SWC users on the mobile app stores if APIs were made available. The product owner decides APIs show the potential to increase the company's reach and grow the business. The question is: what type of APIs should SWC create?

Signs You Need an API

Here are some reasons that an organization may decide to create an API:

- It would like to extend the reach of a core product or service to a broader audience, such as value-add products or onto additional platforms.
- It has have an existing application or system to provide partner access to. For example, a company with a medical billing platform may create APIs to allow medical offices and hospitals to submit invoices.
- People are accessing its website via web scraping or reverse engineering website APIs, which indicates a demand for an API.
- It has valuable data, analytics, or metrics to provide to the public or partners.
- It has created statistical or machine learning models to share.
- It has developed generative AI models to share with application builders.

Selecting the First API Products

The SportsWorldCentral product team performs additional user research by contacting the individuals who have placed help desk tickets and requested features. They also send a survey to their fantasy customers and reach out to the top five fantasy advice websites. They review online message boards and scour social media to find complaints or comments related to their website. And then they conduct some competitive analyses of other fantasy league hosts to see what APIs they provide.

Identifying Potential Users

Based on your research, you narrow your focus down to a few users. Your fantasy manager customers haven't asked for APIs, but they want the things that APIs will bring. They want to import their teams into the fantasy advice websites so that they can get advice on managing their teams. The advice website providers want to import SWC leagues into their websites so that SWC managers will subscribe to their service. Data science users would like to create analytics dashboards, charts, and models using the data from the website, combined with other public sources. Mobile app developers would like to create apps in the mobile app stores to help SWC managers run their teams. Generative AI developers would also like to create chatbots and other applications that can read current SWC data and give advice to managers.

Table 1-1 summarizes these users with the tasks they are trying to perform and their *pain points*, which are problems they face when trying to perform that task.

Table 1-1. Potential consumers for your APIs

User type	Primary tasks	Pain points
Current SWC team managers	Viewing teams and leagues in fantasy advice sites	SWC league host is not supported by advice sites, and no mobile apps are available.
Advice website provider	Giving advice and creating analytics products that support as many league hosts as possible. APIs are the preferred method.	SWC data is not accessible through APIs or any other method.
Data science users	Creating analytics products	They can't access SWC data reliably.
Mobile app developers	Creating third-party apps using SWC data	SWC is not available for use by mobile apps.
Generative AI developers	Creating chat agents and LLM apps using SWC data	No easy access to current SWC data.

Although it is exciting to see so many potential users of APIs, the SWC product manager needs to select the first APIs to create. You work with the product managers to identify the APIs that would satisfy these users, what data would be required to support them, and any changes that would need to be made to the existing website. With that information, the product managers apply a method from design thinking, which is to evaluate the potential products using these three criteria:

- *User desirability*: Your users want the product.
- *Technical feasibility*: Your technical environment and team can create it.
- *Economic viability*: You expect it to be worth the investment.

Creating User Stories

After applying these criteria, they identify user needs that can be fulfilled with existing data without any major changes to the website. To make sure they understand what they'll be building, they create *user stories*, which are informal descriptions of a feature or product that are written from the end user's perspective. A common template for user stories is the following:

- As a (*user type*)
- I want to (*goal or intent*)
- So that (*motivation or benefits*)

The SWC product managers create the following user stories to document the needs they learned their users have:

1. As an *SWC team manager*, I want to *view my fantasy league and team on advice websites* so that *I can win my league and beat my friends.*

2. As an *advice website provider*, I want to *create analytics products such as roster advice, league analysis, and playoff predictions using current-season SWC data* so that *I can increase the number of customers who use my website and increase ad or subscription revenue.*

3. As a *data science user*, I want to *create analytics products such as dashboards, charts, models, and metrics using SWC data* so that *I can demonstrate and grow my data science skills, explore hypotheses and hunches about fantasy data, and build my reputation in both fields.*

4. As a *generative AI developer*, I want to *create AI applications such as chatbots using generative AI and LLMs* so that *I can provide fantasy management advice to SWC fantasy managers.*

Congratulations! The user research you conducted has proven very valuable. You identified several users who need your data and identified ways to serve them that should generate new business for your company. The user stories you captured make an excellent starting point for your API development, which will begin in Chapter 2.

Extending Your Portfolio Project

In addition to the SportsWorldCentral portfolio project, you may want to use similar techniques to build a custom portfolio project in another business domain as you proceed through the book. This is a great way to apply your learning to a data source that you are familiar with or interested in learning. In each chapter, I will suggest some ways that you can apply the techniques you have learned for another portfolio project that is uniquely yours.

Here is how you can extend your project based on this chapter:

- Identify an additional business or market that would benefit from APIs. Perform online research you can use to learn about potential API consumers. Document the user types, primary tasks, and pain points of the users using the user story template.

- Alternative: Research and use the tools from another formal technique such as Design Thinking (*https://oreil.ly/GABFo*), Lean Startup (*https://oreil.ly/w7Alm*), or APIOps Cycles (*https://oreil.ly/apio*) to document your users.

As you continue through the rest of Part I, you can use the techniques you learn about API development to create custom APIs based on these user stories.

Additional Resources

To learn more about API product management, I recommend *APIs: A Strategy Guide* by Daniel Jacobson, Greg Brail, and Dan Woods (O'Reilly, 2011) and *Continuous API Management, 2nd Edition*, by Mehdi Medjaoui, Erik Wilde, Ronnie Mitra, and Mike Amundsen (O'Reilly, 2021).

For more details about design thinking and human-centered design, read the IDEO Field Guide to Human-Centered Design (*https://oreil.ly/GZW5E*).

Summary

You accomplished a lot in this first chapter, and the fun is just starting. Let's review what you have learned so far:

- You learned the tasks that data scientists perform and the tools they use.
- You learned how to design an API that data scientists will love.
- You identified several potential consumers and found two that you could help now: data science users and advice websites.
- You focused on user desirability, technical feasibility, and economic viability to select three user stories to begin your API development.

In Chapter 2, you will start creating an API to fulfill the needs you identified in this chapter. It will, of course, be an API that data scientists and AI will love.

Selecting Your API Architecture

The happy towns are those that have an architecture.
—Le Corbusier, *Towards a New Architecture* (Dover Publications, 1965)

In Chapter 1, you began your portfolio project by understanding user needs and selecting use cases to fulfill. That initial work is critical to make sure you build the right products to fulfill real customer needs. In this chapter, you will begin developing the architecture you will use to build your first APIs.

API Architectural Styles

One of the most significant decisions to make is selecting the API architectural style you will be using. Since you are using a consumer-centric design process, it follows that one of your first goals would be to use a style that is widely supported and understood by potential consumers. The Postman 2023 State of the API Report (*https://oreil.ly/x25Zw*) found these were the top six API architectural styles:

- *REST*: 86%
- *Webhooks*: 36%
- *GraphQL*: 29%
- *Simple Object Access Protocol (SOAP)*: 26%
- *WebSockets*: 25%
- *gRPC*: 11%

The overwhelming popularity of REST found in the survey is consistent with what you will experience if you explore most public APIs. REST is currently the typical style used for APIs. For an example relevant to your project, all of the APIs that I have found for real-world fantasy football league hosts use REST.

There are a couple of other API architectural styles that are worth reviewing because they also make sense in data science and AI-related situations. Let's take a closer look at REST, GraphQL, and gRPC.

Representational State Transfer (REST)

REST was formally defined by Roy Fielding's doctoral dissertation, "Architectural Styles and the Design of Network-Based Software Architectures" (*https://oreil.ly/cLpNu*). In practice, you will find that not all of the REST-style APIs conform completely to this format definition.

A useful implementation of this architectural style is sometimes referred to as *Pragmatic REST* or *RESTful*. The following is a mix of formal definitions and some pragmatic practices:

- API providers make resources available at individual addresses (e.g., */customers*, */products*, etc.). Consumers make requests to these resources using standard HTTP verbs. Producers provide a response. This is the client/server model.
- The response is defined by the producer. The standard structure of the response is the same for each consumer.
- The REST response is typically in JSON or, sometimes, XML format, both of which are standard text-based data transfer formats.
- The interaction is stateless, which means that each message back and forth stands on its own. So, in a conversation of multiple requests and responses, each request has to provide information or *context* from previous responses. For example, a consumer might retrieve a list of players and then provide one player's ID to request additional details.
- Increasingly, REST APIs are defined by an OpenAPI Specification file (*https://oreil.ly/d7xYG*), although a variety of other options have been used over the years.
- It is a best practice to use API versions to protect existing consumers from changes.

Graph Query Language (GraphQL)

GraphQL (*https://oreil.ly/gphql*) is both a query language for APIs and a query run-time engine. GraphQL was developed by Facebook and was made open source in 2015. Here are some attributes of GraphQL APIs, with comparisons to REST:

- Communication uses the client/server model (like RESTful APIs).

- Communication is stateless (like RESTful APIs).

- The response is usually in JSON (like RESTful APIs).

- Instead of only using HTTP verbs, the consumer uses the GraphQL query language.

- The consumer can specify the contents of the response, along with the query options. (In REST, the producer defines the response contents.)

- The producer makes the API available at a single address (e.g., */graphql*), and the consumer passes queries to it via the HTTP POST verb.

- Versioning is not recommended, because the consumer defines the contents they are requesting.

A big advantage of GraphQL over RESTful APIs is that fewer API calls are needed for the consumer to get the information they need. This requires less network traffic.

gRPC

Like GraphQL, gRPC (*https://oreil.ly/iJFdv*) was developed by a commercial company (Google) and was made open source in 2015. gRPC was developed for very fast, efficient communication between microservices. gRPC is usually used for a different set of problems than REST, and it has many differences:

- Instead of sharing resources, gRPC provides remote procedure calls, which are more like traditional code functions.

- Instead of being limited to stateless request-response patterns, gRPC can be used for continuous streaming.

- Instead of returning data in a text-based format like JSON, it uses protocol buffers (*https://protobuf.dev*), which is a format for serializing data that is smaller and faster than JSON or XML.

- Instead of using an OpenAPI specification file, it uses protocol buffers as the specification in a *.proto* file.

gRPC is not a likely candidate for the APIs that you will be creating in your portfolio project. However, it's worth mentioning in this discussion of API architectural styles related to data science for one big reason: *large language models* (LLMs). These machine learning models are the engines behind generative AI services such as Gemini and ChatGPT. These are very big models that need all the performance they can get, and they are using gRPC in some cases to achieve this.

Your Choice: REST

For your company's needs, REST is the appropriate choice. It is the industry standard, and it is appropriate for providing resource-based APIs for the user stories you identified. It is also supported by a broad range of technologies, so your customers should have no problem using a RESTful API.

GraphQL is also a good choice for a user querying your football data, and you should keep an eye on it in the future. But you are safe to wait until your users request it.

Before diving into the Python coding for your API, let's discuss how this book will use a couple of key terms. For this book, we will consider a *RESTful API* to be a set of endpoints that are all related to the same data source. From this perspective, your SWC website will start with a single API: the SWC Fantasy Football API.

An *API version* is a group of endpoints that are consistent for some time so that consumers can count on them.

An *API endpoint* (also referred to as an *operation*) is a combination of two fundamental building blocks: an HTTP verb and a URL path.

The overall structure of these terms is:

```
api
└── version
     └── endpoint
```

Let's look at a few examples using a general Acme widget company (Table 2-1). We'll assume that the company's APIs reside under the *https://api.acme.com* subdomain.

Table 2-1. Example endpoints in Acme API version 1

Endpoint description	HTTP verb	URL	Message body
Read product list	GET	api.acme.com/v1/products/	Empty
Read individual product	GET	api.acme.com/v1/products/{product_id}/	Empty
Create new product	POST	api.acme.com/v1/products/	Contains new product
Update existing product	PUT	api.acme.com/v1/products/{product_id}/	Contains updated product
Delete existing product	DELETE	api.acme.com/v1/products/{product_id}/	Varies

You can see that the URL is reused for several of the endpoints. For example, the read product list endpoint uses a URL of *api.acme.com/v1/products* with a GET verb. The GET verb reads the product records. The create new product endpoint uses the same URL but with a POST verb. The POST creates a new record.

But by combining the HTTP verb with the URL, a specific action is taken when this resource is called. This HTTP verb plus URL combination can only be used once. For your portfolio project, you will develop a set of endpoints to fulfill the user cases you selected.

> For more information about all the HTTP verbs, see "HTTP Basics" on page 167.

Technology Architecture

The SWC league host website is a web application that uses a relational database. Fantasy managers access the website through their web browsers. Although a large amount of technology is necessary to host this large website, Figure 2-1 is a high-level view of the current state application architecture.

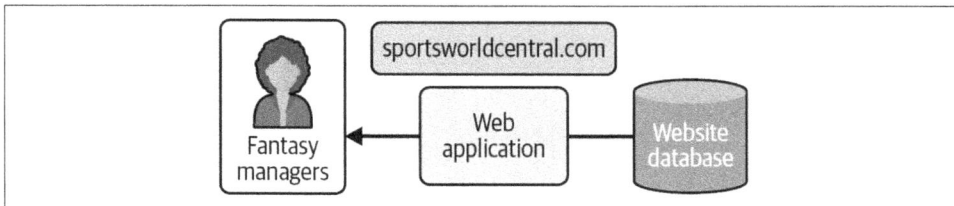

Figure 2-1. Current state application architecture

The following components are in this architecture:

Fantasy managers
    Current web users of Sportsworldcentral.com

Web application
    The existing league host website (assume it already exists for your project)

Website database
    The relational database that is used for the web application (assume it already exists and contains data for your project)

When deciding on the architecture for an API, many choices are available. When a web application already exists, one option is to create the API as part of the web application. Many web applications use APIs as part of their design, so this may be the simplest route in those cases. Another option is to create the API as a separate application but allow the API to read directly from the website database. This has the advantage that the API's data will always be up-to-date with the web application, but it could potentially slow down the web application if a large number of requests are being made to the API.

You will create the API as a separate application and pull the API's data from a *read replica database*. This is a read-only copy of the website database that receives quick updates from the website database but is physically separate so that the API traffic doesn't impact the website's performance. Since your API will be read-only, the read replica is a good choice.

Figure 2-2 shows the future state application architecture that you will implement.

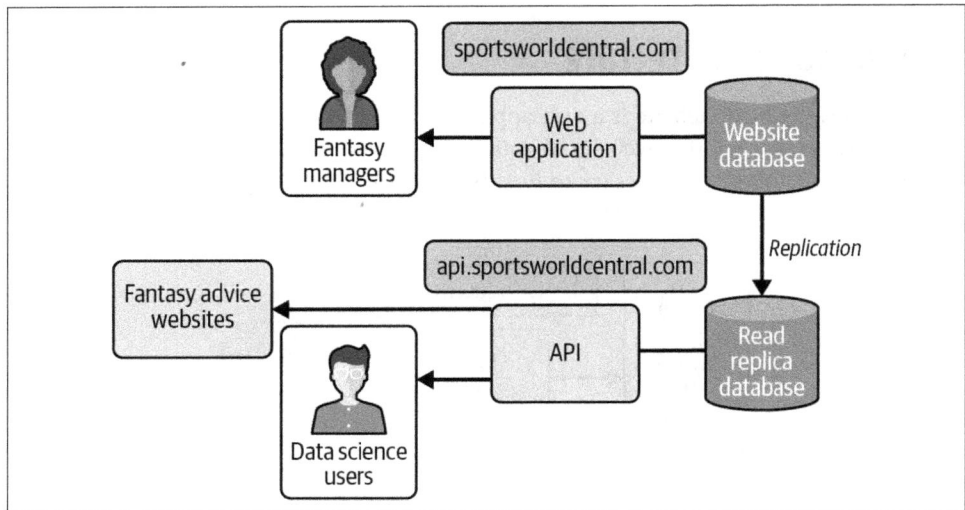

Figure 2-2. Future state application architecture

These are the new components for your project:

Fantasy advice websites
    Will be importing your league data to their advice website.

Data science users
    Will be calling your APIs to create their analytics products.

Read replica database
    A separate read-only copy of the website database you will create.

API
    The new API application you will create. Notice that it will use the base web address *api.sportsworldcentral.com* to keep it separate from the main website.

The technology architecture of APIs is a deep and very interesting topic, and the potential variations of components are nearly endless. One thing to remember is that a software architecture changes over time, so decisions that you make are not permanent. I highly recommend the book *Mastering API Architecture: Design, Operate, and Evolve API-Based Systems* by James Gough, Daniel Bryant, and Matthew Auburn (O'Reilly, 2022) to build foundational knowledge of this subject.

Software Used in This Chapter

Since you will be creating a new API application and standalone database, take a look at the tools and services you will use, as shown in Table 2-2.

Table 2-2. New tools or services used in this chapter

Software name	Version	Purpose
Python	3.10	Programming language
GitHub	NA	Source control, development environment

Python

Python is the programming language you will use for all of the API code in Part I. You will also use it in Part II to create analytics products, data pipelines, and interactive data applications. You will use it in Part III to build a generative AI application. It's possible to use Python for almost any job in data science, and it is the language most frequently used by data scientists, according to the Anaconda 2022 State of Data Science Report (*https://oreil.ly/fD4u7*).

The Python open source ecosystem is very strong and deep, with high-quality frameworks and libraries available for almost any task you want to perform. You will work with a variety of popular Python libraries throughout this book, such as those that follow.

Python adoption has accelerated in recent years for a variety of software development tasks. The 2023 Stack Overflow Developer Survey (*https://oreil.ly/pvfEB*) found that Python was in a dead heat for first as the language most developers wanted to use, and it ranked as the third most used at the time. Python is very flexible and is used in a variety of situations. It is a great tool for any developer to have in their toolbox.

For this book, you will be using Python 3.10 or higher.

GitHub

GitHub is a website that plays a major role in software development. At its core, GitHub is a cloud host of source control software, but it has added additional features over the years. These capabilities are generally free or low cost. Many prominent open source projects use GitHub to host their source code and allow developers to contribute to the project.

You will use GitHub in several ways in this book. You will store all of your program code in repositories while you develop it. You will use GitHub Codespaces as your Python development environment. You will use GitHub Pages to publish your developer portal.

This book uses many of GitHub's tools because they simplify environment management and work together well. The result will be a professional-looking API and data science portfolio that demonstrates what you have accomplished. Most of the work can also be performed on your local machine or another virtual environment instead of using GitHub. However, the instructions will assume you are using GitHub.

Getting Started with Your GitHub Codespace

GitHub Codespaces will be the development environment for all of the API code you develop in Part I of this book. You can think of a Codespace as a development environment running VS Code in the browser. Working with Codespaces will allow you to run the code from the GitHub repository that I share with you, with a minimum of distractions.

Creating Your GitHub Account

Before getting started with Codespaces, you need a GitHub account. Follow the instructions to create a GitHub free personal account (*https://oreil.ly/7j595*). The free account will give you plenty of Codespace storage and core hours to work through this book. During the writing of this book, I often exceeded the free allocation by running multiple Codespaces, but the charge was generally a few dollars. Be sure to enable two-factor authentication to protect your account.

Cloning the Part I Repository

When you use another repository and make edits that you want to keep, there are two ways to accomplish this: forking or cloning. When you *fork* the repository, you create a copy in your account that remains linked to the original repository. This is useful when you will be submitting changes back to the original repository for updates. *Cloning* the repository creates a standalone copy in your GitHub account. For this book, I recommend that you clone the repository so that your portfolio project stands alone and shows the work you have completed.

To clone the repository, log in to GitHub and go to the GitHub Import Repository page (*https://github.com/new/import*). Enter the following information in the fields on this page:

- *The URL for your source repository*: **https://github.com/handsonapibook/api-book-part-one**

- *Your username for your source code repository*: Leave this blank.
- *Your access token or password for your source code repository*: Leave this blank.
- *Repository name*: **portfolio-project**
- *Public*: Select this so that you can share the results of the work you are doing.

Click Begin Import. The import process will begin and the message "Preparing your new repository" will be displayed. After several minutes, you will receive an email notifying you that your import has finished. Follow the link to your new cloned repository.

I will tell you more about the contents of this repository after you launch a GitHub Codespace.

Launching Your GitHub Codespace

Launching a Codespace to work with this repository is simple. On this repository, click the Code button and select the Codespaces tab. Click "Create Codespace on main." You should see a page with the status "Setting up your Codespace." Your Codespace window will open as the setup continues. When the setup completes, your display will look similar to Figure 2-3.

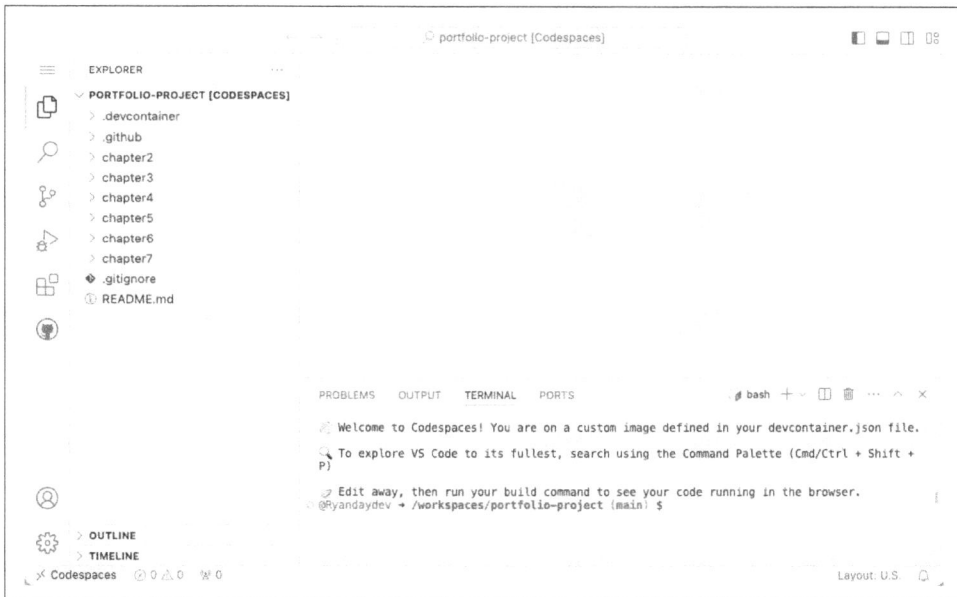

Figure 2-3. New GitHub Codespace

Your Codespace is now created with the cloned repository. This is the environment you will be using for the rest of Part I of this book. Before looking around, take a minute to open the GitHub Codespaces page (*https://github.com/codespaces*) and make a couple of updates. Scroll down the page to find this new Codespace, click the ellipsis to the right of the name, and select Rename. Enter the name **Portfolio project codespace** and click Save. You should see the message "Your codespace *Portfolio project codespace* has been updated." Click the ellipsis again, and then click the ribbon next to "Auto-delete Codespace" to turn off autodeletion.

Touring Your New Codespace

Go back to the tab that has your Codespace open, which looks like Figure 2-3. The URL of this page is an auto-generated address such as *happy-circus-1234.github.dev*. This will be the URL you will use to come back to this Codespace. You could bookmark it if you like. I typically go to the GitHub Codespaces page (*https://github.com/ codespaces*) and launch my Codespace from there.

The display that you see for Codespaces is VS Code, which is a popular open source IDE. Working in Codespaces will be like using VS Code through the browser. Remember that all of the work you do is being executed in a remote container that is running on the cloud. If you would like to know more about how this works, take a look at the GitHub Codespaces overview (*https://oreil.ly/cdesps*).

On the left side of the screen is the Explorer, which shows the directory structure of your Codespace. This structure begins with the structure from the cloned repository. There is a separate subfolder for each chapter in the book that has code. For example, the *chapter2* folder is where you will do your coding for this chapter. Inside each chapter folder is a *complete* subfolder that contains a working copy of the completed code from this chapter.

> I suggest that you follow along with the chapter and type the files yourself. You will learn the purpose of the files as you perform the work. If you run into any trouble, the files in *complete* are available to check your work. If you would like to complete the chapters out of order, you can also use the completed files from the previous chapters as the starting point.

At the bottom of the window, you will see the Terminal window selected. This is an interface to the command line of the Linux container that is running your Code spaces. Throughout Part I, you will enter commands in the terminal window.

Your Codespace has been preloaded with the version of Python that you need. Verify this by entering **python3 --version** in the terminal command line. You should see Python version 3.10 or later, as shown here:

```
$ python3 --version
Python 3.10.13
```

Stop the Codespace by clicking in the bottom left of the window and entering **Stop current codespace** from the dialog window. This will reduce the number of free hours you use working in Codespaces.

Congratulations! This is the repository and Codespace you will use for your Part I portfolio project.

Making Your First Commit

Restart your Codespace from the GitHub Codespaces page (*https://github.com/codespaces*). There is one item to notice as you work in this environment. Your new Codespace begins with the same directory structure as your repository on GitHub.com. However, files you add or make in your Codespace are initially stored only on the Codespace—they are not updated in your repository automatically. As you complete the code in this book and develop your project, you should periodically *commit* your changes to GitHub, which saves your changes and adds a message about the purpose of the change. Frequent commits ensure that you don't lose changes if something happens to your Codespace and allow you to go back to a working copy if you break something. Consistent commits over time demonstrate activity in your GitHub profile, which is a sign of credibility for those viewing your GitHub history.

You do not want to commit everything in your Codespace to your repository. There will be some files that get generated that you don't need to save in GitHub. This is where the *.gitignore* file comes in. Open this file and take a look at it now.

The *.gitignore* file contains file patterns or specific names of all the files in your local Codespace that should be excluded from your repository. Many of these are local config files generated by the libraries you use. Some are sensitive files that should not be published in your repository.

> At a minimum, you should always commit your work when you complete a working session. I commit code several times an hour, when I have completed a chunk of work that is related. For example, if I am modifying multiple files to add a new scoring field to an API, I modify each file and then make a commit with the comment "Added scoring field." (It is not necessary to state the files you changed in the comment, because GitHub tracks that.)

Next, you will update the *README.md* file in the root directory. Each GitHub repository has a README file that provides information about the purpose and contents of the repository. It is written in *Markdown*, which is a lightweight formatting language. In your Codespace, click on the *README.md* file in the Explorer. Modify the text of this file as shown here, and then save it:

```
# API Portfolio Project
This project demonstrates API coding best practices using Python and FastAPI.

This project was built from examples from the book
[Hands-On APIs For API and Data Science](https://hands-on-api-book.com).
```

To preview what this file will look like on GitHub, right-click the *README.md* file and select Open Preview. You will see the updates you made. From the burger menu on the upper left of the window, select File, then Save. Now that you have saved this file, the built-in version control has flagged this. First, you see that the *README.md* file in the Explorer has changed color. Beside that file you see an M, which stands for modified. In the left sidebar, the source control icon has a colored circle with the numeral 1. This means that one change is available to commit to source control. Click on that source control icon, and you will see the Source Control tab, as shown in Figure 2-4.

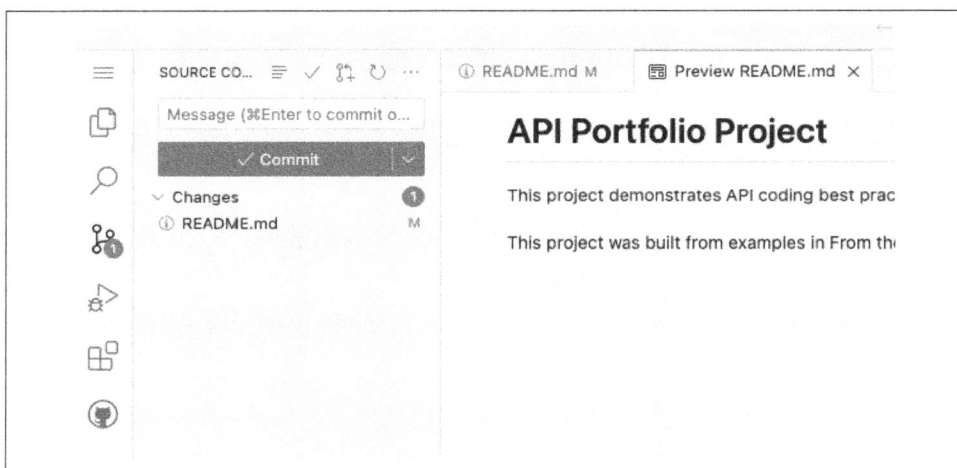

Figure 2-4. Source control with change flagged

Updating your repository requires several steps:

1. Add commit message: The message you add will be the subject of the commit in GitHub.

2. Stage your changes: Identify the changed files to be included in the commit.

3. Commit the changes to your local repository: Your Codespace comes preloaded with a Git repository, and your changes get committed there first.

4. Sync the changes from your local repo to GitHub: Push your changes up to the repository at the GitHub.com website, and pull down any changes that occurred directly on the website.

To save time from here going forward, I'm going to walk you through a shortcut that will save a few steps as you commit in the future. For step 1, add this message in the box: **Personalize title of project.** Next, click the Commit button. A dialog will be displayed that says, "There are no staged changes to commit. Would you like to stage all your changes and commit them directly?" Click Always.

By choosing to always stage all changes and commit them, you are combining steps 2 and 3. For step 4, click Sync Changes to send the updates to the repository at GitHub, which is the *origin* or source of the files. When the dialog is displayed that says "This action will pull and push commits from and to origin/main," click OK, Don't Show Again. One last dialog will be displayed that says "Would you like Visual Studio Code to periodically run git fetch?" Click Yes.

In another tab, go to GitHub.com and open the *portfolio-project* repository, which should look like Figure 2-5.

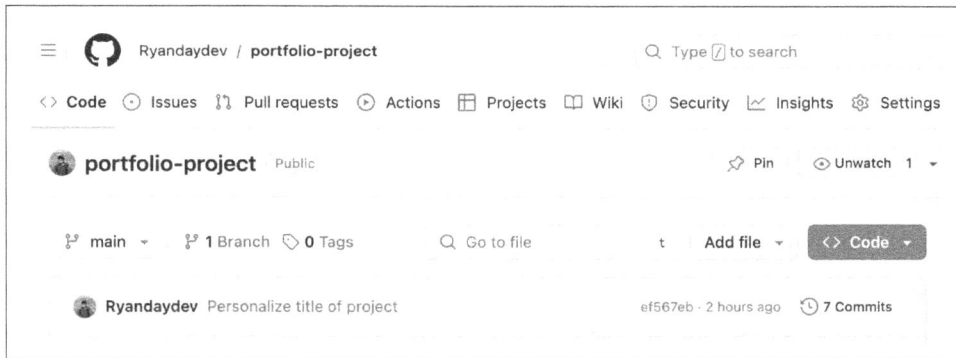

Figure 2-5. Repository showing first commit

Above the directories, click the linked text "Personalize title of your project." This is the record of the commit that you just sent. You can see the files that you changed. When you want to commit changes going forward, you only need to add a title and click Commit, then click Sync Changes to commit your code to GitHub.

This is excellent progress. You have made your first commit, and you are working in your Codespace environment. It's time to get started with your source data.

Additional Resources

To learn about the technical architecture of APIs, I recommend *Mastering API Architecture: Design, Operate, and Evolve API-Based Systems*, by James Gough, Daniel Bryant, and Matthew Auburn (O'Reilly, 2022).

For some tips about RESTful API design, read "The Ten REST Commandments" by Steve McDougall (*https://oreil.ly/_97zo*).

Summary

In this chapter, you started creating an API to fill the user needs you identified in Chapter 1. Here is what you have accomplished so far:

- You defined the API endpoints needed to complete the user stories.
- You set up your development environment using GitHub Codespaces.

In Chapter 3, you will create the website database using SQLite, create Python code to read the database with SQLAlchemy, and perform unit testing with pytest.

Creating Your Database

"You don't seem to give much thought to the matter in hand," I said at last, interrupting Holmes' musical disquisition.

"No data yet," he answered. "It is a capital mistake to theorize before you have all the evidence. It biases the judgement."

"You will have your data soon," I remarked.

    —Sir Arthur Conan Doyle, *A Study in Scarlet* (Ward Lock & Co., 1887)

In Chapter 2, you designed the API architecture and set up your GitHub Codespace environment. In this chapter, you will create the database and the Python code to read from it.

Since you are creating a data API, this chapter is important. It will walk you through the process of designing your database structures, creating them in the SQLite database, creating Python code to read the database, and creating unit tests to verify all of these pieces work together.

If you are in a rush to see how this code works, you can use the files in the *chapter3/ complete* folder, and come back later to follow the instructions step by step.

Components of Your API

In the previous chapter, Figure 2-2 showed the application architecture you are implementing. The API in that diagram is made up of several components. Figure 3-1 shows these components and the software you will use to implement them.

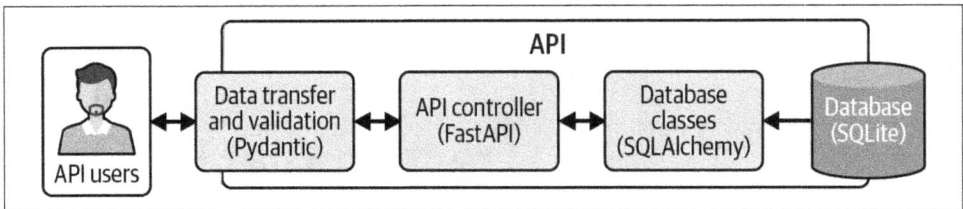

Figure 3-1. API components

There are four major subcomponents of the API. The data transfer and validation components are used to ensure that the API requests and responses have valid data and conform to their definitions. You will create these with Pydantic. You will create the API controller with FastAPI. It handles all of the processing of the API along with other functions you will learn. You will create the database classes using SQLAlchemy. These classes handle querying the database and storing the data in Python classes. Since SQLite is a file-based database and you'll deploy it along with your API code, the diagram shows it as a fourth component of the API.

Software Used in This Chapter

The software introduced in this chapter will focus on databases: creating them, reading data from them, and testing them (see Table 3-1).

Table 3-1. New tools or services used in this chapter

Software name	Version	Purpose
pytest	8	Unit-testing library
SQLAlchemy	2	Object-Relational Mapping (ORM) library to connect Python to SQLite
SQLite	3	Stores the data used by the APIs

SQLite

As shown in Figure 3-1, the API uses the read replica database, which is a read-only copy of the website database that receives quick updates from the website database. The SWC website contains large amounts of data about fantasy teams, NFL players, managers, scoring, and numerous other data points that used by a fantasy football league host.

For your project, you will simulate this with a condensed database using SQLite. SQLite is well suited for learning projects because it is file based and the entire database can easily be stored in a Git repository like the one you'll be using.

Although considered a lightweight database, SQLite supports all the SQL commands that you will use and is fully supported by SQLAlchemy, which you'll use for Python database work. It is a great choice to begin the prototyping of a project. You might replace it with a traditional database such as PostgreSQL or MySQL as the application or API develops. But it is used in many production applications as well.

You will use SQLite 3 for your project.

SQLAlchemy

SQLAlchemy is a popular Python database toolkit and ORM. It works nicely with FastAPI, which will be introduced in Chapter 3. Here are a few of the jobs that SQL-Alchemy does for Python developers:

- It provides query access to databases using Python, without using SQL.
- It populates Python objects with the data from the source database without requiring any conversion of data types.
- It supports a variety of databases.
- It allows the same Python code to be used with different underlying databases.
- It creates queries as prepared statements, which combat SQL injection attacks.

> *SQL injection* is a serious vulnerability in any software that accepts input from users and queries a database with it, including web applications and APIs. It occurs when bad actors insert malicious code into inputs that are intended for data values.
>
> Using *prepared statements* (also known as parameterized statements) instead of raw SQL queries is one technique to reduce the risk of SQL injection. For more information, reference OWASP's article on SQL injection (*https://oreil.ly/24SAy*).

You will be using SQLAlchemy 2 for your project.

pytest

You will be using pytest, a Python testing library, throughout Part I to create tests for the Python code you write. You will create *unit tests* to verify that individual parts of your code work as intended. You will also use it to *regression-test* your code as you make changes or update libraries.

You will be using pytest 8 for your project.

Creating Your SQLite Database

Change to *chapter3* and open SQLite with a new database:

```
.../portfolio-project (main) $ cd chapter3
.../chapter3 (main) $ sqlite3 fantasy_data.db
SQLite version 3.45.3 2024-04-15 13:34:05
Enter ".help" for usage hints.
sqlite>
```

The version of SQLite may differ from what is shown, because it is automatically included in your Codespace.

> To save screen real estate, I have trimmed the directory listing in the terminal prompt of my Codespace. You can configure this by editing the */home/codespace/.bashrc* file in VS Code. Find the export PROMPT_DIRTRIM statement and set it to export PROMPT_DIRTRIM=1. Then, execute this terminal command: **source ~/.bashrc**.

Creating Database Tables

For this project, you will create several tables and load them with data. Figure 3-2 displays the structure of the tables you will create.

You will create these tables by executing Structured Query Language (SQL) statements. As mentioned previously, SQL is a language used frequently by data scientists. This book does not teach the syntax of SQL, but the scripts used are fairly basic. To learn more about SQL, I recommend *Learning SQL: Generate, Manipulate, and Retrieve Data, 3rd Edition*, by Alan Beaulieu (O'Reilly, 2020).

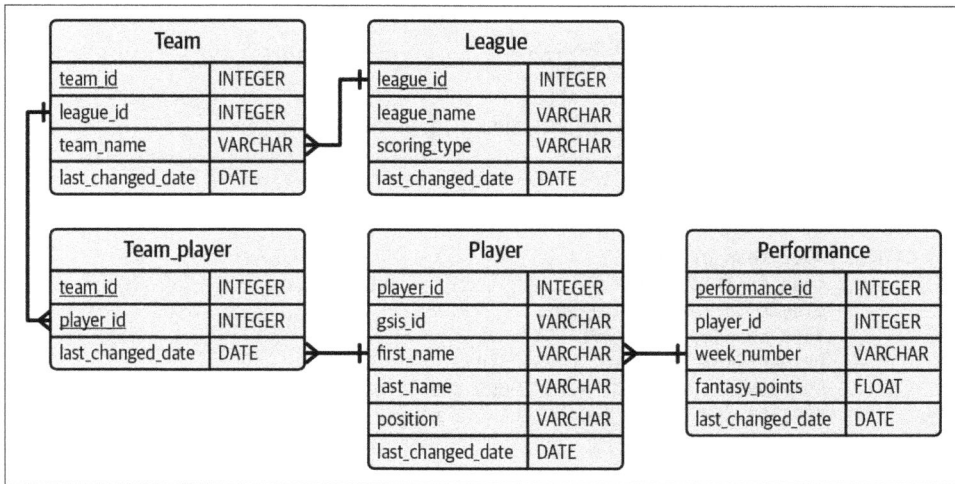

Figure 3-2. Database table structure

Be sure you are still at the `sqlite` prompt, and one-by-one execute the following SQL statements, one by one:

```
CREATE TABLE player (
        player_id INTEGER NOT NULL,
        gsis_id VARCHAR,
        first_name VARCHAR NOT NULL,
        last_name VARCHAR NOT NULL,
        position VARCHAR NOT NULL,
        last_changed_date DATE NOT NULL,
        PRIMARY KEY (player_id)
);

CREATE TABLE performance (
        performance_id INTEGER NOT NULL,
        week_number VARCHAR NOT NULL,
        fantasy_points FLOAT NOT NULL,
        player_id INTEGER NOT NULL,
        last_changed_date DATE NOT NULL,
        PRIMARY KEY (performance_id),
        FOREIGN KEY(player_id) REFERENCES player (player_id)
);

CREATE TABLE league (
        league_id INTEGER NOT NULL,
        league_name VARCHAR NOT NULL,
        scoring_type VARCHAR NOT NULL,
        last_changed_date DATE NOT NULL,
        PRIMARY KEY (league_id)
);
```

```
CREATE TABLE team (
        team_id INTEGER NOT NULL,
        team_name VARCHAR NOT NULL,
        league_id INTEGER NOT NULL,
        last_changed_date DATE NOT NULL,
        PRIMARY KEY (team_id),
        FOREIGN KEY(league_id) REFERENCES league (league_id)
);

CREATE TABLE team_player (
        team_id INTEGER NOT NULL,
        player_id INTEGER NOT NULL,
        last_changed_date DATE NOT NULL,
        PRIMARY KEY (team_id, player_id),
        FOREIGN KEY(team_id) REFERENCES team (team_id),
        FOREIGN KEY(player_id) REFERENCES player (player_id)
);
```

Here is a breakdown of one of the statements:

```
CREATE TABLE player ( ❶
        player_id INTEGER NOT NULL, ❷
        gsis_id VARCHAR,
        first_name VARCHAR NOT NULL,
        last_name VARCHAR NOT NULL,
        position VARCHAR NOT NULL,
        last_changed_date DATE NOT NULL,
        PRIMARY KEY (player_id) ❸
);
```

❶ CREATE TABLE is standard SQL syntax, and player is the name for this table.

❷ The player_id is the name of a single column with a data type of INTEGER that is a required field. If you insert a row in this table with this value, an error will occur.

❸ After all the columns are defined, the player_id value is defined as the *primary key*, which is the value in this table that will always be unique and can be used to join to other tables.

To verify that all five tables were created, enter **.tables**, resulting in the following:

```
sqlite> .tables
league        performance  player      team        team_player
sqlite>
```

Including an External Identifier in Your API

When data scientists use your API, they would like to combine your data with other sources for richer analytics products and models. To accomplish this, you need to provide them a standard external identifier that will be present in other data sources. For NFL player data, each data provider and fantasy website uses a different identifier. The most comprehensive identifier is generated by the NFL's Game Statistics and Information System (GSIS). You will include the GSIS ID in your API to meet the needs of data scientists. Not every player has a GSID ID assigned, so it will allow nulls.

Understanding Table Structure

The first thing to notice about the tables is that each column has a data type defined. The data types used are INTEGER for identifier values, VARCHAR for names and text fields, DATE for date fields, and FLOAT for scoring value fields that have a decimal. All fields have a NOT NULL statement because they are required. Each table has a PRIMARY KEY() constraint on the table's identifier field. This ensures that these values are unique in each table.

In "Designing APIs for Data Scientists" on page 5, I recommended that APIs support querying by the last changed date. This allows them to use APIs in data pipelines and only retrieve records that have changed since a point in time. (This is a major time-saver.) You will enable this by populating the last_changed_date column in each table.

As shown in Figure 3-2, each table is related to at least one other table. This is accomplished using FOREIGN KEY statements in the child table, which references the primary key in the parent table. For an example, look at the definition of the team table:

```
CREATE TABLE team (
        team_id INTEGER NOT NULL,
        team_name VARCHAR NOT NULL,
        league_id INTEGER NOT NULL,
        last_changed_date DATE NOT NULL,
        PRIMARY KEY (team_id),
        FOREIGN KEY(league_id) REFERENCES league (league_id)
);
```

The FOREIGN KEY statement inserts a column named league_id into the child table (team), which links it to a matching record in the parent table (league).

The team_player is the only table that has two foreign keys, as shown:

```
CREATE TABLE team_player (
        team_id INTEGER NOT NULL,
        player_id INTEGER NOT NULL,
```

```
last_changed_date DATE NOT NULL,
PRIMARY KEY (team_id, player_id),
FOREIGN KEY(team_id) REFERENCES team (team_id),
FOREIGN KEY(player_id) REFERENCES player (player_id)
);
```

It has two foreign keys because it is an *association table*, which serves as a child that associates two separate parent tables. In your database, a player can be on many fantasy teams and a team can have many fantasy players. The `team_player` table enables this *many-to-many relationship*. Later, this relationship will be reflected in the Python classes that are mapped to these tables.

The rest of the tables follow a similar design, with data fields that serve the purpose of the data they are storing. You are ready to load data into the tables.

Loading Your Data

Now that the tables are created, you will populate them with football data. You will use SQLite's `.import` tool to load data files that are in CSV format. You will find data files in this chapter's */data* directory.

Before you execute the import, you need to configure SQLite to enforce foreign keys. This means that if you try to insert a record into a child table (a table containing the `FOREIGN KEY` statement) that doesn't match a record in the parent table (the table named in the `REFERENCES` of a foreign key), an error will occur and the record won't be imported. For example, when foreign keys are enforced, you can't insert a record in the `performance` table that uses a `player_id` value that isn't in the `player` table.

Turn on foreign key enforcement with the following statement:

```
sqlite> PRAGMA foreign_keys = ON;
```

Prepare the import statement to recognize CSV format with the following command:

```
sqlite> .mode csv
```

Run the following commands from the `sqlite` prompt to load the data. Run them in the order shown here:

```
sqlite> PRAGMA foreign_keys = ON;
sqlite> .mode csv
sqlite> .import --skip 1 data/player_data.csv player
sqlite> .import --skip 1 data/performance_data.csv performance
sqlite> .import --skip 1 data/league_data.csv league
sqlite> .import --skip 1 data/team_data.csv team
sqlite> .import --skip 1 data/team_player_data.csv team_player
sqlite>
```

Use the following commands to verify that the correct number of records was loaded into each table. The `performance` table has been loaded with records using two different `last_changed_date` values so that you can verify date searching functions are working correctly:

```
sqlite> select count(*) from player;
1018
sqlite> select count(*) from performance;
17306
sqlite> select count(*) from performance where last_changed_date > '2024-04-01';
2711
sqlite> select count(*) from league;
5
sqlite> select count(*) from team;
20
sqlite> select count(*) from team_player;
140
```

To exit the SQLite application, type **.exit**:

```
sqlite> .exit
$
```

You have loaded sample data in your database, which represents the data from the SWC website data. Now you'll start using it with Python.

Accessing Your Data Using Python

There are several ways to access this data in Python. For example, you could create a connection to the database and execute SQL queries directly. This sounds simple, but you would quickly run into several issues, such as mapping the SQLite data types into Python objects. You would also need to take steps to avoid SQL injection attacks when you accept input from your API users.

To avoid this manual work, you will use an ORM, which handles the process of reading database tables and creating Python objects from them. You will be using a very common Python ORM: SQLAlchemy.

Installing SQLAlchemy in Your Environment

SQLAlchemy is the first Python library that you will need to install directly in your Codespace. You want to be certain of the version of SQLAlchemy installed, so first create a `pip` requirements file in the directory with your Python code.

In your editor, create a file named *requirements.txt* with the following contents, and save the file:

```
SQLAlchemy>=2.0.0
```

This file will be used to install libraries from the `pip` Python package manager. These libraries are stored on the internet, and `pip` will automatically download them to your Codespace.

Throughout the book, you will add additional Python libraries to your Codespace. Using the requirements file is a convenient way to install multiple libraries and make sure the versions of the libraries are all compatible with one another.

To install the library, execute the following command:

```
pip3 install -r requirements.txt
```

You should see a message that says SQLAlchemy 2.0 or higher has been successfully installed or was "already satisfied." To verify the installation, type **pip3 show SQL Alchemy** and you will receive a confirmation similar to the following:

```
Name: SQLAlchemy
Version: 2.0.29
Summary: Database Abstraction Library
Home-page: https://www.sqlalchemy.org
Author: Mike Bayer
Author-email: mike_mp@zzzcomputing.com
License: MIT
```

Selecting and Updating Open Source Library Versions

When you use open source libraries, you need to plan and test which versions of each library to use. Because many libraries have dependencies on each other, it can be tricky at times to find library versions that all work together. In addition, you have to frequently review the versions of each library when patches are needed or versions stop being maintained. How can you verify your code will still work if you update a library? Two key items will help: virtual environments and regression tests.

Codespaces are an ideal virtual environment for testing. You can create a new Codespace from your code repository, uninstall the existing libraries, and reinstall the new ones using the *requirements.txt* file.

You will create basic unit tests in this chapter and Chapter 4 to verify your database and API code. Run these tests in the new virtual environment with the new libraries and see if any errors or warnings occur that require coding changes. This is *regression testing*: finding what you broke when updating your code or libraries. The more code you cover with tests, the more confident you can be when updating libraries.

In general, using a library with a major version of 1 or higher indicates the library will be maintained and patched for a while when a newer version is released. Keep in mind that open source libraries generally are provided without any warranty from the volunteer maintainers.

Creating Python Files for Database Access

You will now create the files that are required to query the database using Python. Table 3-2 explains the purpose of all the files you will have when this chapter is complete.

Table 3-2. Purpose of the Chapter 3 files

Filename	Purpose
crud.py	Helper function to query the database
database.py	Configures SQLAlchemy to use the SQLite database
models.py	Defines the SQLAlchemy classes related to the database tables
requirements.txt	Used to install specific versions of libraries with the `pip` package manager
test_crud.py	The pytest file to unit-test your SQLAlchemy files

The file named *models.py* will contain the Python representation of the data. The classes in this file will be used when you query databases in Python.

Here are the two tasks that you need to perform in this file:

- Define the SQLAlchemy classes to store information from database tables.
- Describe the relationship between these tables so that the Python code can access the related tables.

> The term *model* is used in a lot of different ways in this book, which is unavoidable but confusing. In this instance, the SQL-Alchemy model is a Python representation of the data from the SQLite database.

Create a file with the following contents, and name it *models.py*:

```python
"""SQLAlchemy models"""
from sqlalchemy import Column, ForeignKey, Integer, String, Float, Date
from sqlalchemy.orm import relationship

from database import Base


class Player(Base):
    __tablename__ = "player"

    player_id = Column(Integer, primary_key=True, index=True)
    gsis_id = Column(String, nullable=True)
    first_name = Column(String, nullable=False)
    last_name = Column(String, nullable=False)
```

```python
    position = Column(String, nullable=False)
    last_changed_date = Column(Date, nullable=False)

    performances = relationship("Performance", back_populates="player")


    # Many-to-many relationship between Player and Team tables
    teams = relationship("Team", secondary="team_player",
                         back_populates="players")


class Performance(Base):
    __tablename__ = "performance"

    performance_id = Column(Integer, primary_key=True, index=True)
    week_number = Column(String, nullable=False)
    fantasy_points = Column(Float, nullable=False)
    last_changed_date = Column(Date, nullable=False)

    player_id = Column(Integer, ForeignKey("player.player_id"))

    player = relationship("Player", back_populates="performances")


class League(Base):
    __tablename__ = "league"

    league_id = Column(Integer, primary_key=True, index=True)
    league_name = Column(String, nullable=False)
    scoring_type = Column(String, nullable=False)
    last_changed_date = Column(Date, nullable=False)

    teams = relationship("Team", back_populates="league")


class Team(Base):
    __tablename__ = "team"

    team_id = Column(Integer, primary_key=True, index=True)
    team_name = Column(String, nullable=False)
    last_changed_date = Column(Date, nullable=False)

    league_id = Column(Integer, ForeignKey("league.league_id"))

    league = relationship("League", back_populates="teams")

    players = relationship("Player", secondary="team_player",
                           back_populates="teams")

class TeamPlayer(Base):
    __tablename__ = "team_player"
```

```
        team_id = Column(Integer, ForeignKey("team.team_id"),
                         primary_key=True, index=True)
        player_id = Column(Integer, ForeignKey("player.player_id"),
                           primary_key=True, index=True)
    last_changed_date = Column(Date, nullable=False)
```

Take a look at *models.py* piece by piece. At the top of most Python files, you will find `import` statements. The power of the Python ecosystem comes from the variety of external libraries you can use. The process you will use in this book is to install the libraries using the `pip` package manager, and then reference them in your code using import statements:

```
from sqlalchemy import Column, ForeignKey, Integer, String, Float, Date ❶
from sqlalchemy.orm import relationship ❷

from database import Base ❸
```

❶ Because this file will create Python representations of the database tables, you first import the data types that SQLAlchemy will use for the database fields. For more information about SQLAlchemy data types, reference the SQLAlchemy Type Hierarchy (*https://oreil.ly/Z1jfo*).

❷ Next, you import SQLAlchemy's relationship functionality, which enables foreign key relationships between tables.

❸ The `database` import refers to the *database.py* file with the SQLAlchemy configuration. You are using the `Base` class, which is a standard template you'll use for the classes in the *models.py* file.

Now it's time to begin the definition of the `Player` class, which is the Python class you'll use to store data from the SQLite `player` table. You do this using the `class` statement, stating the name of the class and specifying that it will be a subclass of the `Base` template imported from the *database.py* file. Use the magic command *table name* to tell SQLAlchemy to reference the `player` table. Because of this statement, when you ask SQLAlchemy to query `Player`, it will know behind the scenes to access the `player` table in the database. This is one of the key benefits of an ORM—mapping the Python code automatically to the underlying database:

```
class Player(Base):
    __tablename__ = "player"
```

The rest of the `Player` class definition maps additional details about the database table. Each statement defines one attribute in the class using the `Column` method provided by SQLAlchemy:

```
        player_id = Column(Integer, primary_key=True, index=True)
        gsis_id = Column(String, nullable=True)
        first_name = Column(String, nullable=False)
```

```
last_name = Column(String, nullable=False)
position = Column(String, nullable=False)
last_changed_date = Column(Date, nullable=False)
```

Here are a few things to notice about the definitions:

- The attribute names are automatically matched to the column names in the database.
- The data types used (e.g., `String`, `Integer`) are SQLAlchemy data types that you specified in your `import` statement at the beginning of the file.
- The `primary_key` definition provides several benefits from SQLAlchemy, such as query optimization, enforcing uniqueness, and enabling relationships between classes.

Along with the definition of the tables, you define the foreign key relationship between the tables using the `relationship()` function. This results in a `Player` `.performances` attribute that will return all the related rows from the `performance` table for each row in the `player` table:

```
performances = relationship("Performance", back_populates="player")
```

There is another kind of relationship, which uses the `team_player` association table to connect `player` to `team`. By defining `secondary="team_player"`, this relationship allows a `Player` record to have an attribute named `Player.teams`. This is the many-to-many relationship that was discussed when creating the database tables:

```
players = relationship("Player", secondary="team_player",
                       back_populates="teams")
```

Next is the definition for the `Performance` class:

```
class Performance(Base):
    __tablename__ = "performance"

    performance_id = Column(Integer, primary_key=True, index=True)
    week_number = Column(String, nullable=False)
    fantasy_points = Column(Float, nullable=False)
    last_changed_date = Column(Date, nullable=False)

    player_id = Column(Integer, ForeignKey("player.player_id"))

    player = relationship("Player", back_populates="performances")
```

This class has a `player` relationship that is the mirror image of the `performances` relationship in the `player` table. When you look at these two relationships together, you can see that the `back_populates` statement in one refers to the variable assigned in the other. Together these allow a two-way relationship between the parent (`player`) and child (`performance`).

Next up is the League class:

```
class League(Base):
    __tablename__ = "league"

    league_id = Column(Integer, primary_key=True, index=True)
    league_name = Column(String, nullable=False)
    scoring_type = Column(String, nullable=False)
    last_changed_date = Column(Date, nullable=False)

    teams = relationship("Team", back_populates="league")
```

League is going to be the topmost parent class in your code, as was reflected in Figure 3-2. The teams relationship will be used to enable League.teams in this class and has a matching relationship in the Team class.

Look at the next block of code, which defines the Team class:

```
class Team(Base):
    __tablename__ = "team"

    team_id = Column(Integer, primary_key=True, index=True)
    team_name = Column(String, nullable=False)

    league_id = Column(Integer, ForeignKey("league.league_id"))

    league = relationship("League", back_populates="teams")

    players = relationship("Player", secondary="team_player",
                           back_populates="teams")
```

Notice that this file has matching relationships to connect with the league table and indirectly to the player table.

The last class definition is for the team-player table:

```
class TeamPlayer(Base):
    __tablename__ = "team_player"

    team_id = Column(Integer, ForeignKey("team.team_id"),
                     primary_key=True, index=True)
    player_id = Column(Integer, ForeignKey("player.player_id"),
                       primary_key=True, index=True
```

The TeamPlayer class is created without any relationships, because those are defined on the Team and Player classes. You have now defined all of the SQLAlchemy models needed for the new database tables and the necessary database configuration file. Excellent progress!

Creating the Database Configuration File

Next, a file named *database.py* will set up the SQLAlchemy configuration to connect to the SQLite database, along with some other Python objects that you'll use for database work. The tasks that you need to accomplish in this file are the following:

- Create a database connection that points to the SQLite database and has the correct settings.
- Create a parent class that you'll use to define the Python table classes:

Create a file with the following contents, and name it *database.py*:

```python
"""Database configuration"""
from sqlalchemy import create_engine
from sqlalchemy.orm import declarative_base
from sqlalchemy.orm import sessionmaker

SQLALCHEMY_DATABASE_URL = "sqlite:///./fantasy_data.db"

engine = create_engine(
    SQLALCHEMY_DATABASE_URL, connect_args={"check_same_thread": False}
)
SessionLocal = sessionmaker(autocommit=False, autoflush=False, bind=engine)

Base = declarative_base()
```

Take a look at this file piece by piece. Three specific functions are imported from the SQLAlchemy libraries. Although it would be possible to import the entire SQLAlchemy library all at once, it is better to import specific functions to limit possible conflicts between duplicate functions in multiple libraries:

```python
from sqlalchemy import create_engine
from sqlalchemy.orm import declarative_base
from sqlalchemy.orm import sessionmaker
```

The next three steps work together to get the *session*, which is a SQLAlchemy object that manages the conversation with the database. Create a database URL that tells SQLAlchemy what type of database you'll be using (SQLite) and where to find the file (in the same folder as this file, with the name *fantasy_data.db*):

```python
SQLALCHEMY_DATABASE_URL = "sqlite:///./fantasy_data.db"
```

Using this database URL, create an engine object, with one configuration setting that allows multiple connections to this database without an error being thrown:

```python
engine = create_engine(
    SQLALCHEMY_DATABASE_URL, connect_args={"check_same_thread": False}
)
```

Then, use the `engine` object to create a session named `SessionLocal` that points to that engine and adds a couple of more configuration settings:

```
SessionLocal = sessionmaker(autocommit=False, autoflush=False, bind=engine)
```

The last command in this file creates a `Base` class. This is a standard template SQL-Alchemy provides for the models you will create in the *models.py* file:

```
Base = declarative_base()
```

Creating SQLAlchemy Helper Functions

The files created so far give you a connection to the database and classes that represent database tables. Next, you will create the file *crud.py* that contains query functions. This strange-sounding name stands for Create, Read, Update, Delete (CRUD).

Create a file with the following contents, and name it *crud.py*:

```python
"""SQLAlchemy Query Functions"""
from sqlalchemy.orm import Session
from sqlalchemy.orm import joinedload
from datetime import date


import models


def get_player(db: Session, player_id: int):
    return db.query(models.Player).filter(
        models.Player.player_id == player_id).first()


def get_players(db: Session, skip: int = 0, limit: int = 100,
                min_last_changed_date: date = None,
                last_name : str = None, first_name : str = None, ):
    query = db.query(models.Player)
    if min_last_changed_date:
        query = query.filter(
            models.Player.last_changed_date >= min_last_changed_date)
    if first_name:
        query = query.filter(models.Player.first_name == first_name)
    if last_name:
        query = query.filter(models.Player.last_name == last_name)
    return query.offset(skip).limit(limit).all()


def get_performances(db: Session, skip: int = 0, limit: int = 100,
                     min_last_changed_date: date = None):
    query = db.query(models.Performance)
    if min_last_changed_date:
        query = query.filter(
            models.Performance.last_changed_date >= min_last_changed_date)
```

```python
        return query.offset(skip).limit(limit).all()


def get_league(db: Session, league_id: int = None):
    return db.query(models.League).filter(
        models.League.league_id == league_id).first()


def get_leagues(db: Session, skip: int = 0, limit: int = 100,
                min_last_changed_date: date = None,league_name: str = None):
    query = db.query(models.League
                     ).options(joinedload(models.League.teams))
    if min_last_changed_date:
        query = query.filter(
            models.League.last_changed_date >= min_last_changed_date)
    if league_name:
        query = query.filter(models.League.league_name == league_name)
    return query.offset(skip).limit(limit).all()


def get_teams(db: Session, skip: int = 0, limit: int = 100,
              min_last_changed_date: date = None,
              team_name: str = None, league_id: int = None):
    query = db.query(models.Team)
    if min_last_changed_date:
        query = query.filter(
            models.Team.last_changed_date >= min_last_changed_date)
    if team_name:
        query = query.filter(models.Team.team_name == team_name)
    if league_id:
        query = query.filter(models.Team.league_id == league_id)
    return query.offset(skip).limit(limit).all()


#analytics queries
def get_player_count(db: Session):
    query = db.query(models.Player)
    return query.count()


def get_team_count(db: Session):
    query = db.query(models.Team)
    return query.count()


def get_league_count(db: Session):
    query = db.query(models.League)
    return query.count()
```

Let's look at the import statements in *crud.py*:

```
from sqlalchemy.orm import Session ❶
from sqlalchemy.orm import joinedload
from datetime import date ❷

import models ❸
```

❶ Session and joinedload are used by the query functions.

❷ The date will be an important data type to allow you to filter by date.

❸ This import lets you reference the model file that you created. These functions reference the classes that you created in *models.py* and use SQLAlchemy built-in functions to retrieve data using prepared SQL statements.

Take a look at the first query:

```
def get_player(db: Session, player_id: int):
    return db.query(models.Player).filter(
        models.Player.player_id == player_id).first()
```

The parameters in this function include a database session, which the function will use to connect to the database, and a specific player_id value. By using filter(models.Player.player_id == player_id).first(), this function looks up a specific Player.player_id value and returns the first matching instance. Because you have defined player_id as a primary key in the *models.py* file and the SQLite database, this query will return a single result.

The signature of the next function adds several new parameters to the .query() statement:

```
def get_players(db: Session, skip: int = 0, limit: int = 100,
                min_last_changed_date: date = None,
                last_name : str = None, first_name : str = None, ):
```

The skip and limit parameters will be used for *pagination*, which allows the user to specify a set of records in chunks rather than a full list. The min_last_changed_date parameter will be used to exclude records older than a specified date.

The int = 0 on the skip parameter sets a default value of zero. If this parameter isn't sent in a call to this function, skip will default to zero. The limit has a default of 100. There is no default given for min_last_changed_date, first_name, and last_name, so those default to null.

The body of the function uses the queries to filter the results:

```
query = db.query(models.Player)
    if min_last_changed_date:
        query = query.filter(
            models.Player.last_changed_date >= min_last_changed_date)
    if first_name:
        query = query.filter(models.Player.first_name == first_name)
    if last_name:
        query = query.filter(models.Player.last_name == last_name)
```

The last statement applies the skip and limit parameters:

```
    return query.offset(skip).limit(limit).all()
```

This statement applies the skip and limit parameters to grab a specific chunk of records from the query results. The skip instructs the query to skip a number of records from the beginning of the results, and limit instructs the query to return only a certain number of records. For instance, a user might begin by skipping zero and limiting 20. This would return the first 20 records. They could call it again, this time skipping 20 and limiting 20. This would grab the next 20.

The get_leagues function uses a new statement, so it is worth a closer look:

```
def get_leagues(db: Session, skip: int = 0, limit: int = 100,
                min_last_changed_date: date = None,league_name: str = None):
    query = db.query(models.League
                    ).options(joinedload(models.League.teams))
    if min_last_changed_date:
        query = query.filter(
            models.League.last_changed_date >= min_last_changed_date)
    if league_name:
        query = query.filter(models.League.league_name == league_name)
    return query.offset(skip).limit(limit).all()
```

This function uses the .options(joinedload(models.League.teams)) statement. This is a type of *eager loading*, which causes SQLAlchemy to retrieve the joined team data when it retrieves the league data.

The final set of queries are designed to support AI and large language models, based on the recommendation to provide a separate endpoint for analytics questions. You will create endpoints that provide counts for users, leagues, and teams. This will help the AI use the pagination functions, and it will answer questions about the number of records without making large API calls:

```
#analytics queries
def get_player_count(db: Session):
    query = db.query(models.Player)
    return query.count()

def get_team_count(db: Session):
    query = db.query(models.Team)
```

```
        return query.count()

def get_league_count(db: Session):
    query = db.query(models.League)
    return query.count()
```

You have created all the SQLAlchemy classes and helper functions. Since all of the functions in *crud.py* are reading (querying) data, you have only implemented the "r" in CRUD. That is appropriate, because all of your user stories require read-only functionality. If you were developing an API that allowed creating, updating, or deleting records, this file could be extended with additional functions. Now it is time to unit-test these queries with pytest.

Installing pytest in Your Environment

Now that all the database code is written, you are ready to test it. You will use the pytest library for this task. First, add an entry to the *requirements.txt* file for pytest. The updated file should look like the following:

```
SQLAlchemy>=2.0.0
Pytest>=8.1.0
```

To install pytest, execute the following command again:

pip3 install -r requirements.txt

You should see a message that says pytest 8.1.0 or higher has been successfully installed or was "already satisfied." To verify the installation, type **pip3 show Pytest** and you will receive a confirmation similar to the following:

```
$ pip3 show Pytest
Name: pytest
Version: 8.1.1
Summary: pytest: simple powerful testing with Python
Home-page:
Author: Holger Krekel, Bruno Oliveira, Ronny Pfannschmidt, Floris Bruynooghe,
Brianna Laugher, Florian Bruhin, Others (See AUTHORS)
Author-email:
License: MIT
```

Testing Your SQLAchemy Code

As the library's summary says, pytest is simple to use. There are a couple of naming conventions that pytest expects. Any file that contains tests will have a filename beginning with *test* followed by an underscore or ending with an underscore followed by *test*. Inside the test file, pytest will execute any function name beginning with *test*.

Inside the test functions, you will include an `assert` statement. If it returns true, the flow continues. If all assertions evaluate as true in the test, the test returns with a success status. If an assertion evaluates as false, the code raises an `AssertionError` and the test evaluates as false.

Your unit tests will be very basic: they will check that the row counts returned from your SQLAlchemy classes match the values you checked in the previous SQL query.

Create a file named *test_crud.py* with the following contents:

```python
"""Testing SQLAlchemy Helper Functions"""
import pytest
from datetime import date

import crud
from database import SessionLocal

# use a test date of 4/1/2024 to test the min_last_changed_date.
test_date = date(2024,4,1)

@pytest.fixture(scope="function")
def db_session():
    """This starts a database session and closes it when done"""
    session = SessionLocal()
    yield session
    session.close()

def test_get_player(db_session):
    """Tests you can get the first player"""
    player = crud.get_player(db_session, player_id = 1001)
    assert player.player_id == 1001

def test_get_players(db_session):
    """Tests that the count of players in the database is what is expected"""
    players = crud.get_players(db_session, skip=0, limit=10000,
                               min_last_changed_date=test_date)
    assert len(players) == 1018

def test_get_players_by_name(db_session):
    """Tests that the count of players in the database is what is expected"""
    players = crud.get_players(db_session, first_name="Bryce", last_name="Young")
    assert len(players) == 1
    assert players[0].player_id == 2009


def test_get_all_performances(db_session):
  """Tests that the count of performances in the database is
    what is expected - all the performances"""
    performances = crud.get_performances(db_session, skip=0, limit=18000)
    assert len(performances) == 17306

def test_get_new_performances(db_session):
  """Tests that the count of performances in the database is
    what is expected"""
```

```
performances = crud.get_performances(db_session, skip=0, limit=18000,
                                      min_last_changed_date=test_date)
```

```
#test the count functions
def test_get_player_count(db_session):
    player_count = crud.get_player_count(db_session)
    assert player_count == 1018
```

First, look at how this file follows the conventions expected by pytest. The file is named *test_crud.py*, so it will be recognized as a test file automatically. The file contains six function names beginning with *test_*. These will be executed when the file runs. Each of these test functions ends with an `assert` statement.

The first function needs a bit of explanation. On top of the function is the decorator `@pytest.fixture(scope="function")`. A fixture is used during the *arrange* phase, which prepares the testing setup. This fixture uses session scope, which means it will run once for each function:

```
@pytest.fixture(scope="function")
```

The body of the `db_session()` function creates a database session, pauses while the test function uses the session (through the `yield` statement), and then closes the session when the test completes:

```
def db_session():
    """This starts a database session and closes it when done"""
    session = SessionLocal()
    yield session
    session.close()
```

To verify the date-based queries are working correctly, the queries for `performance` check the full results and then results that are limited using `last_changed_date`. First remember that in the SQL queries earlier you got the following results for the `performance` table:

```
sqlite> select count(*) from performance;
17306
sqlite> select count(*) from performance where last_changed_date > '2024-04-01';
2711
```

To verify the first result using pytest, this function does not include a data parameter:

```
def test_get_all_performances(db_session):
    """Tests that the count of performances in the database is
    what is expected - all the performances"""
    performances = crud.get_performances(db_session, skip=0, limit=18000)
    assert len(performances) == 17306
```

To verify the second result, the next function uses a `last_changed_date` value of 2024_04_01, set in the `test_date` variable at the top of the testing code. That date is earlier that all by 2,711 records:

```
"""Tests that the count of performances in the database is
what is expected"""
performances = crud.get_performances(db_session, skip=0, limit=10000,
                                     min_last_changed_date=test_date)
assert len(performances) == 2711
```

The last test verifies one of the analytics queries:

```
def test_get_player_count(db_session):
    player_count = crud.get_player_count(db_session)
    assert player_count == 1018
```

To execute the tests, enter the **pytest test_crud.py** command and you should see an output that looks similar to this:

```
$ pytest test_crud.py
==================== test session starts =====================
platform linux -- Python 3.10.13, pytest-8.1.2, pluggy-1.5.0
rootdir: /workspaces/adding-more-data/chapter3
plugins: anyio-4.4.0
collected 5 items


test_crud.py                                        [100%]

==================== 5 passed in 0.22s =======================
```

You have verified that your SQLAlchemy classes and a few helper functions work correctly—way to go! The database work is done.

Additional Resources

SQL is one of the essential skills for data professionals. The number of resources available is limitless, but here are a couple to start:

- *Learning SQL, 3rd Edition*, by Alan Beaulieu (O'Reilly, 2020)
- *SQL Pocket Guide, 4th Edition*, by Alice Zhao (O'Reilly, 2021)

To learn more about SQLAlchemy, check out the official SQLAlchemy 2 documentation (*https://oreil.ly/PhsUf*).

Summary

In this chapter, you created your database and the SQLAlchemy code to read it. Here is what you accomplished in this chapter:

- You designed your database tables and their relationships.
- You created a database using SQLite and created all of your tables using SQL commands.
- You imported data from CSV files to load your tables.
- You created the Python model files and database configuration files.
- You created helper functions to query your database.
- You unit-tested the end-to-end database functionality using pytest.

In Chapter 4, you will create the FastAPI code to use this data and publish it as a REST API.

Developing the FastAPI Code

In Chapter 3, you created your database and the Python code to access the database. In this chapter, you will build on this foundation code to create a working API. Table 4-1 lists the endpoints that you will create to fulfill these user stories.

Table 4-1. Endpoints for the SWC Fantasy Football API

Endpoint description	HTTP verb	URL
API health check	GET	/
Read player list	GET	/v0/players/
Read individual player	GET	/v0/players/{player_id}/
Read performance list	GET	/v0/performances/
Read league list	GET	/v0/leagues/
Read individual league	GET	/v0/leagues/{league_id}/
Read team list	GET	/v0/teams/
Read counts	GET	/v0/counts/

You are using version 0 for your API. This will notify API consumers that the product is changing rapidly and they should be aware of potential *breaking changes*—changes that cause functionality to stop working and may require consumers to make changes in their program code.

Continuing Your Portfolio Project

Figure 4-1 shows the same API components you saw previously, with one addition: the Uvicorn web server. Uvicorn will execute your API code and interact with API requests.

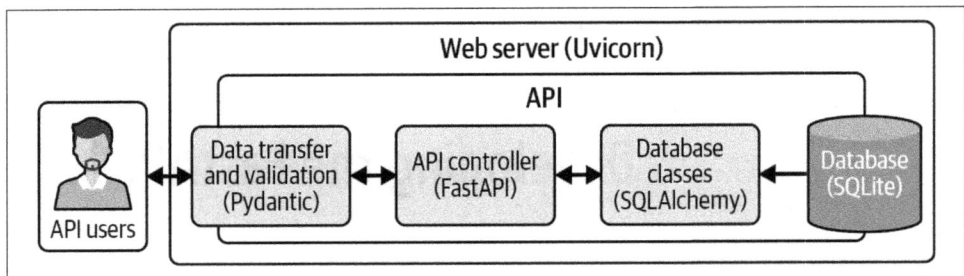

Figure 4-1. API components with Uvicorn

In Chapter 3, you completed two very important parts of the API: the SQLite database and the SQLAlchemy classes that enable Python to interact with the data. In this chapter, you will finish the rest of the components. You will create Pydantic *schemas* that define the structure of request and response messages. Then, you will create the controlling FastAPI application that stitches all the other components together to finish the API.

Software Used in This Chapter

The software introduced in this chapter will focus on handling API requests from your consumers. Table 4-2 lists the new tools you will use.

Table 4-2. New tools used in this chapter

Software name	Version	Purpose
FastAPI	0	Web framework to build the API
FastAPI CLI	0	Command-line interface for FastAPI
HTTPX	0	HTTP client for Python
Pydantic	2	Validation library
Uvicorn	0	Web server to run the API

FastAPI

FastAPI is a Python web framework that is designed for building APIs. A *web framework* is a set of libraries that simplify common tasks for web applications. Other common web frameworks include Express, Flask, Django, and Ruby on Rails.

FastAPI is built to be fast in both application performance and developer productivity. Because FastAPI focuses on API development, it simplifies several tasks related to API building and publishing:

- It handles HTTP traffic, requests/responses, and other "plumbing" jobs with a few lines of code.

- It automatically generates an OpenAPI specification file for your API, which is useful for integrating with other products.
- It includes interactive documentation for your API.
- It supports API versioning, security, and many other capabilities.

As you will see as you work through the portfolio project, all of these capabilities provide benefits to the users of your APIs.

Compared to the other frameworks I mentioned, FastAPI is a relative newcomer. It is an open source project created by Sebastián Ramírez Montaño in 2018.

FastAPI also includes the FastAPI CLI. This is a separate Python library that is used to run FastAPI from the command line.

As of this writing, the latest version of FastAPI is a 0.x version (e.g., 0.115). That version number is important because, according to semantic versioning, 0.x indicates that breaking changes may occur with the software.

HTTPX

HTTPX is a Python HTTP client. It is similar to the very popular requests library, but it supports *asynchronous calls*, which allows some tasks to finish while others process. The requests library only supports *synchronous calls*, which wait until they receive a response before continuing. HTTPX is used by pytest to test FastAPI programs. You will also use this library in Chapter 7 to create your Python SDK.

Pydantic

Pydantic is a data validation library, which will play a key part in the APIs that you build. Because APIs are used to communicate between systems, a critical piece of their functionality is the validation of inputs and outputs. API developers and data scientists typically spend a significant amount of time writing the code to check the data types and validate values that go into and out of the API endpoints.

Pydantic is purpose-built to address this important task. Pydantic is fast in two ways: it saves the developer time that would be spent to write custom Python validation code, and Pydantic validation code runs much faster because it is implemented in the Rust programming language.

In addition to these benefits, objects defined in Pydantic automatically support tooltips and hints in IDEs such as VS Code. FastAPI uses Pydantic to generate JSON Schema representations from Python code. *JSON Schema* is a standard that ensures consistency in JSON data structures. This Pydantic feature enables FastAPI to automatically generate the *OpenAPI specification*, which is an industry-standard file describing APIs.

For your project, you will use Pydantic version 2.

Uvicorn

All web applications, including APIs, rely on a web server to handle the various administrative tasks related to handling requests and responses. You will be using the open source Uvicorn web server. Uvicorn is based on the *ASGI specification*, which provides support for both *synchronous processes* (which block the process while waiting for a task to be performed) and *asynchronous processes* (which can allow another process to continue while they are waiting).

For your project, you will be using Uvicorn 0.x.

You Can Start from Here

The instructions in this chapter assume that you completed Chapters 2 and 3 already. If you're starting your coding in this chapter, you will need to perform a couple of steps to catch up. First, you need to create a GitHub Codespace from the book's GitHub repository. Full instructions are available in "Getting Started with Your GitHub Codespace" on page 22.

To catch up on the coding from Chapter 3, you can use the completed set of files that is in *chapter3/complete* of your Codespace. If you are using these, use the directory *chapter3/complete* instead of *chapter3* in the setup commands that follow.

If you run into trouble with any of the steps in this chapter, there are a few troubleshooting tips at the end.

Copying Files from Chapter 3

To continue your portfolio project where you left it in the previous chapter, change the directory to *chapter4* and then copy the previous chapter's files over to it. The following shows the commands and expected output:

```
.../portfolio-project (main) $ cd chapter4
.../chapter4 (main) $ cp ../chapter3/*.py .
.../chapter4 (main) $ cp ../chapter3/fantasy_data.db .
.../chapter4 (main) $ cp ../chapter3/requirements.txt .
.../chapter4 (main) $ ls *.*
crud.py database.py fantasy_data.db models.py readme.md requirements.txt
test_crud.py
```

Installing the New Libraries in Your Codespace

In the previous chapter, you created the *requirements.txt* file and specified libraries to install using the `pip3` package manager in Python. You will now use this process to install Pydantic, FastAPI, and Uvicorn.

Update *requirements.txt* to match the following:

```
#Chapter 4 pip requirements
SQLAlchemy>=2.0.0
pydantic>=2.4.0
fastapi[standard]>=0.115.0
uvicorn>=0.23.0
Pytest>=8.1.0
httpx>=0.27.0
```

Execute the following command to install the new libraries in your Codespace and verify that the libraries installed in the previous chapter still exist:

```
pip3 install -r requirements.txt
```

You should see a message that states that these libraries were successfully installed, such as the following:

```
Installing collected packages: uvicorn, pydantic, httpx, fastapi
Successfully installed fastapi-0.115.4 httpx-0.26.0 pydantic-2.4.2 uvicorn-0.23.2
```

Creating Python Files for Your API

You will be creating two new Python files, which are detailed in Table 4-3.

Table 4-3. Purpose of the Chapter 4 files

Filename	Purpose
main.py	FastAPI file that defines routes and controls API
schemas.py	Defines the Pydantic classes that validate data sent to the API
test_main.py	The pytest file for the FastAPI program

Creating Pydantic Schemas

The Pydantic classes define the structure of the data that the consumer will receive in their API responses. This uses a software design pattern called *data transfer objects* (DTO), in which you define a format for transferring data between a producer and consumer, without the consumer needing to know the backend format. In your portfolio project, the backend and frontend classes won't look significantly different, but using DTOs allows complete flexibility on this point.

Although you define the classes using Python code and your code interacts with them as fully formed Python objects, the consumer will receive them in an HTTP request as a JSON object. FastAPI uses Pydantic to perform the *serialization* process, which is converting the Python objects into JSON for the API response. This means you do not need to manage serialization in your Python code, which simplifies your program. Pydantic 2 is written in Rust and performs this task much faster than Python could. In addition to performing this de-serialization task, Python also defines the response format in the *openapi.json* file. This is a standard contract that uses OpenAPI and JSON Schema. This will provide multiple benefits for the consumer, as you will see in subsequent chapters. Pydantic will take data from SQLAlchemy classes and provide it to the API users.

> Both SQLAlchemy and Pydantic documentation refer to their classes as models, which may be confusing at times. This is extra confusing for data science work, where models have additional meanings. For clarity, this book will refer to Pydantic schemas and SQLAlchemy models.

Create a file with the following contents, and name it *schemas.py*:

```python
"""Pydantic schemas"""
from pydantic import BaseModel, ConfigDict
from typing import List
from datetime import date


class Performance(BaseModel):
    model_config = ConfigDict(from_attributes = True)
    performance_id : int
    player_id : int
    week_number : str
    fantasy_points : float
    last_changed_date : date

class PlayerBase(BaseModel):
    model_config = ConfigDict(from_attributes = True)
    player_id : int
    gsis_id: str
    first_name : str
    last_name : str
    position : str
    last_changed_date : date

class Player(PlayerBase):
    model_config = ConfigDict(from_attributes = True)
    performances: List[Performance] = []

class TeamBase(BaseModel):
```

```
    model_config = ConfigDict(from_attributes = True)
    league_id : int
    team_id : int
    team_name : str
    last_changed_date : date

class Team(TeamBase):
    model_config = ConfigDict(from_attributes = True)
    players: List[PlayerBase] = []

class League(BaseModel):
    model_config = ConfigDict(from_attributes = True)
    league_id : int
    league_name : str
    scoring_type : str
    last_changed_date : date
    teams: List[TeamBase] = []

class Counts(BaseModel):
    league_count : int
    team_count : int
    player_count : int
```

The schemas in this file will be used to form the responses to the API endpoints that you will define next. The primary schemas are directly returned to the endpoints and the secondary schemas are returned as an attribute of the primary schema. For example, the */v0/players/* endpoint URL returns a list of Player objects (primary), which has the attribute Player.performances (secondary). Table 4-4 shows the mapping between API endpoints and schemas.

Table 4-4. Mapping of schemas to endpoints

Endpoint URL	Primary schema	Secondary schema
/	None	None
/v0/players/	Player	Performance
/v0/players/{player_id}/	Player	Performance
/v0/performances/	Performance	None
/v0/leagues/	League	TeamBase
/v0/leagues/{league_id}	League	TeamBase
/v0/teams/	Team	PlayerBase
/v0/counts/	Counts	None

The Performance class is the first and simplest schema:

```
class Performance(BaseModel):
    model_config = ConfigDict(from_attributes = True)
    performance_id : int
    player_id : int
```

```
week_number : str
fantasy_points : float
last_changed_date : date
```

This class represents the scoring data that the consumer will receive. From their perspective, a *performance* is what happens when a player plays in a single week. If you compare the elements of this class to the SQLAlchemy models, you will see that it contains all of the elements that the `Performance` model contains.

`Performance` is a subclass of the Pydantic `BaseModel` class, which provides a lot of built-in capabilities, including validating the data types, converting the Python object to JSON (serializing), raising intelligent errors, and connecting automatically to the SQLAlchemy models.

> Notice that the Pydantic data types of individual class elements are assigned with a colon, and not an equals sign which is what SQL-Alchemy uses. (This will trip you up if you're not careful.)

The player data is represented in two schemas: `PlayerBase` and `Player`. Breaking the data into two classes allows you to share a limited version of the data in some situations and a full version in others. Here are those two schemas:

```
class PlayerBase(BaseModel):
    model_config = ConfigDict(from_attributes = True)
    player_id : int
    gsis_id: str
    first_name : str
    last_name : str
    position : str
    last_changed_date : date


class Player(PlayerBase):
    model_config = ConfigDict(from_attributes = True)
    performances: List[Performance] = []
```

The performance data had a single `Performance` schema, but the player data has two schemas. `PlayerBase` is a subclass of `BaseModel`, and it has all the player fields except one: the `Performance` list. Table 4-4 shows that `PlayerBase` will be used as a secondary schema for the */v0/teams/* endpoint. The reason is simple: to reduce the amount of data transmitted in the API call. When the API user retries a list of `Team` schemas, they want to see all the players on that team without also getting a list of all the scoring performances for all the players.

The full `Player` schema is a subclass of `PlayerBase` and adds the list of `Performance` objects. This schema is used directly in the */v0/players/* and */v0/players/{player_id}/* endpoints. In those situations, the API user wants a list of scoring performances with the players.

To see the secondary use of `PlayerBase`, examine the next two schemas:

```
class TeamBase(BaseModel):
    model_config = ConfigDict(from_attributes = True)
    league_id : int
    team_id : int
    team_name : str
    last_changed_date : date

class Team(TeamBase):
    model_config = ConfigDict(from_attributes = True)
    players: List[PlayerBase] = []
```

The `Team` object contains the statement `players: List[PlayerBase] = []`. As mentioned previously, this means the items in `Team.players` are of the more limited `PlayerBase` schema. This is the secondary usage of `PlayerBase` shown in Table 4-4 in the */v0/teams/* endpoint.

The next class is the `League` schema:

```
class League(BaseModel):
    model_config = ConfigDict(from_attributes = True)
    league_id : int
    league_name : str
    scoring_type : str
    last_changed_date : date
    teams: List[TeamBase] = []
```

By now you probably noticed that `League.teams` contains `TeamBase` objects. This is the secondary use of `TeamBase` used in the */v0/leagues/* endpoint.

Finally, you will create a special-purpose schema to support the analytics provided by the *v0/counts/* endpoint. This schema does not directly map to a database table, so it does not include the `model_config` element. The name of the schema is `Counts`, and it includes the number of league, team, and player records in the API:

```
class Counts(BaseModel):
    league_count : int
    team_count : int
    player_count : int
```

At this point, you have designed the DTOs that will be used to send data to the API consumer. You are ready for the final piece: the FastAPI controller class.

Creating Your FastAPI Controller

Now that all of the pieces are in place in the other Python files, you can tie them together with the FastAPI functionality in *main.py*. You can accomplish a lot with only a few lines of FastAPI code.

Create the file with the following contents, and name it *main.py*:

```python
"""FastAPI program - Chapter 4"""
from fastapi import Depends, FastAPI, HTTPException
from sqlalchemy.orm import Session
from datetime import date

import crud, schemas
from database import SessionLocal

app = FastAPI()

# Dependency
def get_db():
    db = SessionLocal()
    try:
        yield db
    finally:
        db.close()

@app.get("/")
async def root():
    return {"message": "API health check successful"}


@app.get("/v0/players/", response_model=list[schemas.Player])
def read_players(skip: int = 0,
                 limit: int = 100,
                 minimum_last_changed_date: date = None,
                 first_name: str = None,
                 last_name: str = None,
                 db: Session = Depends(get_db)
                 ):
    players = crud.get_players(db,
                skip=skip,
                limit=limit,
                min_last_changed_date=minimum_last_changed_date,
                first_name=first_name,
                last_name=last_name)
    return players


@app.get("/v0/players/{player_id}", response_model=schemas.Player)
def read_player(player_id: int,
                db: Session = Depends(get_db)):
    player = crud.get_player(db,
```

```
                               player_id=player_id)
    if player is None:
        raise HTTPException(status_code=404,
                            detail="Player not found")
    return player


@app.get("/v0/performances/",
         response_model=list[schemas.Performance])
def read_performances(skip: int = 0,
                limit: int = 100,
                minimum_last_changed_date: date = None,
                db: Session = Depends(get_db)):
    performances = crud.get_performances(db,
                skip=skip,
                limit=limit,
                min_last_changed_date=minimum_last_changed_date)
    return performances


@app.get("/v0/leagues/{league_id}", response_model=schemas.League)
def read_league(league_id: int,db: Session = Depends(get_db)):
    league = crud.get_league(db, league_id = league_id)
    if league is None:
        raise HTTPException(status_code=404, detail="League not found")
    return league



@app.get("/v0/leagues/", response_model=list[schemas.League])
def read_leagues(skip: int = 0,
                limit: int = 100,
                minimum_last_changed_date: date = None,
                league_name: str = None,
                db: Session = Depends(get_db)):
    leagues = crud.get_leagues(db,
                skip=skip,
                limit=limit,
                min_last_changed_date=minimum_last_changed_date,
                league_name=league_name)
    return leagues


@app.get("/v0/teams/", response_model=list[schemas.Team])
def read_teams(skip: int = 0,
                limit: int = 100,
                minimum_last_changed_date: date = None,
                team_name: str = None,
                league_id: int = None,
                db: Session = Depends(get_db)):
    teams = crud.get_teams(db,
                skip=skip,
                limit=limit,
                min_last_changed_date=minimum_last_changed_date,
                team_name=team_name,
                league_id=league_id)
```

```
        return teams


@app.get("/v0/counts/", response_model=schemas.Counts)
def get_count(db: Session = Depends(get_db)):
    counts = schemas.Counts(
        league_count = crud.get_league_count(db),
        team_count = crud.get_team_count(db),
        player_count = crud.get_player_count(db))
    return counts
```

Let's walk through the code in your FastAPI file. We'll begin with the imports:

```
from fastapi import Depends, FastAPI, HTTPException ❶
from sqlalchemy.orm import Session ❷
from datetime import date ❸

import crud, schemas ❹
from database import SessionLocal ❺
```

❶ These are methods from the FastAPI library. You will use these to identify this program as a FastAPI application.

❷ The SQLAlchemy `Session` will be used when this program calls *crud.py*.

❸ You will use the `date` type to query by last changed date.

❹ These imports allow the FastAPI application to reference the SQLAlchemy and Pydantic classes.

❺ This retrieves the shared `SessionLocal` class that is used to connect to your SQLite database.

Continue reviewing the code:

```
app = FastAPI()

# Dependency
def get_db():
    db = SessionLocal()
    try:
        yield db
    finally:
        db.close()
```

In FastAPI, the primary class you will work with is a `FastAPI` class. This class by default includes the functionality to handle much of the work that an API needs to perform, without requiring you to specify every detail. You create a `FastAPI` instance and name it `app`. This will be used in the rest of *main.py*. When you execute your API from the command line using Uvicorn, you will reference `main:app`, referring to the `app` object in *main.py*.

You define the `get_db()` function to create a database session and close the session when you are done with it. This function is used as a dependency in the API routes within *main.py*:

```
@app.get("/")
async def root():
    return {"message": "API health check successful"}
```

The next command is `@app.get("/")`, which is a *decorator*. A decorator is a statement that is added above a function definition, to give special attributes to it. In this case, the decorator defines that the `async def root()` function definition will be a FastAPI request handler.

This function will be called when a consumer accesses the root URL of the API, which is equivalent to /. It will serve as a health check for the entire API by returning a simple message to the consumer. The next statement defines the first endpoint that we have created for your user stories:

```
@app.get("/v0/players/", response_model=list[schemas.Player])
def read_players(skip: int = 0,
                 limit: int = 100,
                 minimum_last_changed_date: date = None,
                 first_name: str = None,
                 last_name: str = None,
                 db: Session = Depends(get_db)
                 ):
    players = crud.get_players(db,
                skip=skip,
                limit=limit,
                min_last_changed_date=minimum_last_changed_date,
                first_name=first_name,
                last_name=last_name)
    return players
```

Remember that Table 4-1 defined the endpoints that we planned to create as a combination of HTTP verb and URL. With FastAPI these endpoints (also called *routes*) are defined with the decorators above each function.

The following explains how the HTTP verb and URL are specified in the decorator:

- *HTTP verb*: All of these endpoints use the `GET` verb, which is defined by the `@app.get()` decorator function.
- *URL*: The first parameter of the `get()` function is the relative URL. For this first endpoint, the URL is */v0/players/*.

The second parameter of the decorator is `response_model=list[schemas.Player])`. This informs FastAPI that the data returned from this endpoint will be a list of Pydantic `Player` objects, as defined in the *schemas.py* file. This information will be included in the OpenAPI specification that FastAPI automatically creates for this API. Consumers can count on the returned data being valid according to this definition.

Let's look at the function signature that you decorated:

```
def read_players(skip: int = 0,
                 limit: int = 100,
                 minimum_last_changed_date: date = None,
                 first_name: str = None,
                 last_name: str = None,
                 db: Session = Depends(get_db)
                 ):
```

Several things are going on in this function. Starting at the end, the db object is a session that is created by the `get_db()` function defined at the top of this file. By wrapping the function in `Depends()`, FastAPI handles the call for and gives the `Session` to your function.

The next two parameters are optional integers with a default value: `skip: int = 0`, `limit: int = 100`, `last_name`. These are followed by two optional string parameters that default to `None`. These are all named parameters that have a defined data type and a default value. FastAPI will automatically include these parameters as query parameters in the API definition. Query parameters are included in the URL path with a question mark in front and an ampersand between.

For instance, to call this query method, the API consumer could use this request:

- *HTTP verb*: GET
- *URL*: *{base URL}/v0/players/?first_name=Bryce&last_name=Young*

Within the body of the `read_players()` function, FastAPI is calling the `get_play ers()` function that you defined in *crud.py*. It is performing a database query. The `players` object receives the result of that function call. FastAPI validates that this object matches the definition `list[schemas.Player]`. If it does, FastAPI uses Pydantic to serialize the Python objects into a text JSON string and sends the response to the consumer.

The next endpoint adds two additional FastAPI features:

```
@app.get("/v0/players/{player_id}", response_model=schemas.Player)
def read_player(player_id: int,
                db: Session = Depends(get_db)):
    player = crud.get_player(db,
                             player_id=player_id)
```

```
    if player is None:
        raise HTTPException(status_code=404,
                            detail="Player not found")
    return player
```

First, the URL path includes {player_id}. This is a *path parameter*, which is an API request parameter that is included in the URL path instead of being separated by question marks and ampersands, like the query parameters. Here is an example of how the API consumer might call this endpoint:

- *HTTP verb*: GET
- *URL*: *{base URL}/v0/players/12345?skip=10&limit=50*

The function checks to see if any records were returned from the helper function, and if not, it raises an *HTTPException*. This is a standard method that web applications use to communicate status. It is good RESTful API design to use the standard HTTP status codes (*https://oreil.ly/cTnfI*) to communicate with consumers. This makes the operation more predictable and reliable. This endpoint returns an HTTP status code of 404, which is the *not found* code. It adds the additional message that the item not found was the player being searched for.

The next four endpoints do not use any new features. But together they complete all of the user stories that we have included for your first API:

```
@app.get("/v0/performances/",
         response_model=list[schemas.Performance])
def read_performances(skip: int = 0,
                limit: int = 100,
                minimum_last_changed_date: date = None,
                db: Session = Depends(get_db)):
    performances = crud.get_performances(db,
                skip=skip,
                limit=limit,
                min_last_changed_date=minimum_last_changed_date)
    return performances

@app.get("/v0/leagues/{league_id}", response_model=schemas.League)
def read_league(league_id: int,db: Session = Depends(get_db)):
    league = crud.get_league(db, league_id = league_id)
    if league is None:
        raise HTTPException(status_code=404, detail="League not found")
    return league


@app.get("/v0/leagues/", response_model=list[schemas.League])
def read_leagues(skip: int = 0,
                limit: int = 100,
                minimum_last_changed_date: date = None,
                league_name: str = None,
                db: Session = Depends(get_db)):
```

```
        leagues = crud.get_leagues(db,
                    skip=skip,
                    limit=limit,
                    min_last_changed_date=minimum_last_changed_date,
                    league_name=league_name)
        return leagues

@app.get("/v0/teams/", response_model=list[schemas.Team])
def read_teams(skip: int = 0,
                limit: int = 100,
                minimum_last_changed_date: date = None,
                team_name: str = None,
                league_id: int = None,
                db: Session = Depends(get_db)):
    teams = crud.get_teams(db,
                skip=skip,
                limit=limit,
                min_last_changed_date=minimum_last_changed_date,
                team_name=team_name,
                league_id=league_id)
        return teams
```

The final endpoint provides counts of leagues, teams, and players:

```
@app.get("/v0/counts/", response_model=schemas.Counts)
def get_count(db: Session = Depends(get_db)):
    counts = schemas.Counts(
        league_count = crud.get_league_count(db),
        team_count = crud.get_team_count(db),
        player_count = crud.get_player_count(db))
    return counts
```

It is worth noting that, in addition to the basic options of FastAPI and Pydantic that you are using, many other validations and features are available. As you can see, these libraries accomplish a lot with only a few lines of code from you.

Testing Your API

You will use pytest to test your *main.py* file. As with the *crud.py* file in the previous chapter, you will be testing that the correct number of records are returned by each API endpoint. The counts of records can be verified by the SQL queries in "Loading Your Data" on page 36.

To implement the tests for your API, create a file with the following contents, and name it *test_main.py*:

```
from fastapi.testclient import TestClient
from main import app

client = TestClient(app)
```

```python
# test the health check endpoint
def test_read_main():
    response = client.get("/")
    assert response.status_code == 200
    assert response.json() == {"message": "API health check successful"}


# test /v0/players/
def test_read_players():
    response = client.get("/v0/players/?skip=0&limit=10000")
    assert response.status_code == 200
    assert len(response.json()) == 1018


def test_read_players_by_name():
    response = client.get("/v0/players/?first_name=Bryce&last_name=Young")
    assert response.status_code == 200
    assert len(response.json()) == 1
    assert response.json()[0].get("player_id") == 2009


# test /v0/players/{player_id}/
def test_read_players_with_id():
    response = client.get("/v0/players/1001/")
    assert response.status_code == 200
    assert response.json().get("player_id") == 1001


# test /v0/performances/
def test_read_performances():
    response = client.get("/v0/performances/?skip=0&limit=20000")
    assert response.status_code == 200
    assert len(response.json()) == 17306


# test /v0/performances/ with changed date
def test_read_performances_by_date():
    response = client.get(
        "/v0/performances/?skip=0&limit=20000&minimum_last_changed_date=
        2024-04-01"
    )
    assert response.status_code == 200
    assert len(response.json()) == 2711


# test /v0/leagues/{league_id}/
def test_read_leagues_with_id():
    response = client.get("/v0/leagues/5002/")
    assert response.status_code == 200
    assert len(response.json()["teams"]) == 8


# test /v0/leagues/
def test_read_leagues():
    response = client.get("/v0/leagues/?skip=0&limit=500")
    assert response.status_code == 200
    assert len(response.json()) == 5


# test /v0/teams/
```

```
def test_read_teams():
    response = client.get("/v0/teams/?skip=0&limit=500")
    assert response.status_code == 200
    assert len(response.json()) == 20

# test /v0/teams/
def test_read_teams_for_one_league():
    response = client.get("/v0/teams/?skip=0&limit=500&league_id=5001")
    assert response.status_code == 200
    assert len(response.json()) == 12

# test the count functions
def test_counts():
    response = client.get("/v0/counts/")
    response_data = response.json()
    assert response.status_code == 200
    assert response_data["league_count"] == 5
    assert response_data["team_count"] == 20
    assert response_data["player_count"] == 1018
```

The file begins with import statements and creation of the TestClient class:

```
from fastapi.testclient import TestClient ❶
from main import app ❷

client = TestClient(app) ❸
```

❶   TestClient is a special class that allows the FastAPI program to be tested without running it on a web server.

❷   This references the FastAPI object you created in *main.py*.

❸   This statement creates a TestClient that will test your application.

Take a look at a few of the test functions:

```
#test the health check endpoint
def test_read_main():
    response = client.get("/")
    assert response.status_code == 200
    assert response.json() == {"message": "API health check successful"}
```

This function uses the TestClient to simulate an API call to the root path. Then, it checks the HTTP status code for a value of 200, which means a successful request. Next, it looks at the JSON value returned by the API and checks that it matches the JSON value provided.

The next test function adds more functionality:

```
#test /v0/players/
def test_read_players():
    response = client.get("/v0/players/?skip=0&limit=10000")
    assert response.status_code == 200
    assert len(response.json()) == 1018
```

Notice that the URL passed in the `get()` statement uses the `skip` and `limit` parameters. The second `assert` statement checks the length of the list of players returned by the API to make sure it is exactly 1018.

Another test function tests the search of players by name. Although the database does not enforce uniqueness on player names, duplicate player names are rare, and names are commonly used to identify players.

This search without a key supports the design recommended in Chapter 1 for AI:

```
def test_read_players_by_name():
    response = client.get("/v0/players/?first_name=Bryce&last_name=Young")
    assert response.status_code == 200
    assert len(response.json()) == 1
    assert response.json()[0].get("player_id") == 2009
```

This adds two `assert` statements: one to make sure only one record was returned from this query (after all, there is only one Bryce Young) and another to make sure the `player_id` is correct.

The complete file contains 11 tests in all. To execute the tests, enter the following command:

```
.../chapter4 (main) $ pytest test_main.py
=================== test session starts ============================
platform linux -- Python 3.10.14, pytest-8.1.2, pluggy-1.4.0
rootdir: /workspaces/portfolio-project/chapter4
plugins: anyio-3.4.4.0

collected 11 items

test_main.py                                        [100%]

=================== 11 passed in 1.01s =============================
```

You have verified that your FastAPI program works with pytest. Now it's time to try it with a web server.

Launching Your API

This is the moment you have been waiting for: it's time to run your API. Enter the following command from the command line:

```
.../chapter4 (main) $ fastapi run main.py
```

You will see the application startup occur as shown in Figure 4-2.

```
 module    2. main.py

   code    Importing the FastAPI app object from the module with the following code:

           from main import app

    app    Using import string: main:app

 server    Server started at http://0.0.0.0:8000
 server    Documentation at http://0.0.0.0:8000/docs

           Logs:

   INFO    Started server process [11779]
   INFO    Waiting for application startup.
   INFO    Application startup complete.
   INFO    Uvicorn running on http://0.0.0.0:8000 (Press CTRL+C to quit)
```

Figure 4-2. FastAPI running from the command line

In Codespaces, you will also see a dialog stating "Your application running on port 8000 is available," as shown in Figure 4-3.

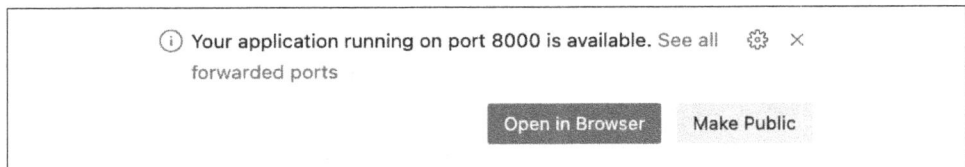

Figure 4-3. Codespaces browser window pop-up

Click "Open in Browser" to open a browser tab outside your Codespaces. This browser will show a base URL ending in *app.github.dev* that contains the response from your API running on Codespaces. You should see the following health check message in your web browser:

```
{"message":"API health check successful"}
```

This confirms your API is running, which is a great start.

The next test is to call an endpoint that retrieves data. Give that a try by copying and pasting the following onto the end of the base URL in your browser: */v0/performances/?skip=0&limit=1*. For example, the full URL might be *https://happy-pine-tree-1234-8000.app.github.dev/v0/performances/?skip=0&limit=1*.

If everything is working correctly, you should see the following data in your browser:

```
[{"performance_id":2501,"player_id":1001,"week_number":"202301",
"fantasy_points":20.0,"last_changed_date":"2024-03-01"}]
```

This chapter covered a lot, so it's possible that an error occurred or you are not getting a successful result. Don't worry, this happens to all of us. Here are a few suggestions for how to troubleshoot any problems you are running into:

- Run the `pip3 install -r requirements.txt` command again to make sure you have all the updated software.
- Take a minute to verify the path in the URL bar of your browser. Minor things matter, such as slashes and question marks.
- Look at the command line to see any errors that are being thrown by FastAPI.
- To verify your environment with FastAPI and Uvicorn, try creating a simple API, such as one from the official FastAPI tutorial (*https://oreil.ly/L7QWz*).
- If a formatting error occurs due to text wrapping, check against the files in the GitHub repository.

If this first API endpoint is working for you, try out some more of the URLs from Table 4-1 in your browser to verify that you have completed all of your user stories. Congratulations, you are an API developer!

Additional Resources

To explore FastAPI beyond this book, the official FastAPI tutorial (*https://oreil.ly/SFN3w*) and FastAPI reference documentation (*https://oreil.ly/MVgVk*) are both very useful.

To learn the ins and outs of building a project with FastAPI, I recommend *FastAPI: Modern Python Web Development* by Bill Lubanovic (O'Reilly, 2023).

For a growing list of practical tips from an official FastAPI Expert, check out Marcelo Trylesinski's FastAPI Tips (*https://oreil.ly/kludex*).

The official Pydantic 2.4 documentation (*https://oreil.ly/2OE-8*) provides information for the specific version of Pydantic used in this chapter.

The official Uvicorn documentation (*https://oreil.ly/uvicorn*) has much more information about the capabilities of this software.

Summary

In this chapter, you completed the API functionality for the SWC Fantasy Football API. You accomplished the following:

- You installed FastAPI, SQLAlchemy, Pydantic, and Uvicorn, along with several supporting libraries.
- You defined Pydantic schemas to represent the data that your API consumers wanted to receive.
- You created a FastAPI program to process consumer requests and return data responses, tying everything together.
- You tested the API with pytest and then ran it successfully on the web server.

In Chapter 5, you will document your API using FastAPI's built-in capabilities.

Documenting Your API

> *Documentation is where developers come to learn about your API. Whether you're providing a simple README file or developing a full website for your developers, it is critical to clearly document how they can most effectively use your API.*
>
> —Brenda Jin, Saurabh Sahni, and Amir Shevat, *Designing Web APIs* (O'Reilly, 2018)

In Chapter 4, you created your first API to fulfill the user stories identified in Chapter 1. The primary users you developed these for are data science users and advice website providers. The API will enable them to create analytics products such as dashboards, charts, and models using the SportsWorldCentral (SWC) fantasy data. Now you will create documentation to help them use your API. Documentation and features such as software development kits (SDKs) improve the *developer experience* (DX) for your technical users.

Sending a Signal of Trust

An important job of your API docs is to signal trust to potential API consumers. If they don't trust you, they won't use your API. If API consumers are going to call your API multiple times, they will build an *integration*, which is basically code that calls your code repeatedly. For the SWC API, this might be a fantasy advice website that will be adding SWC to its list of supported websites. Another example would be a data science user who is going to schedule an Extract, Transform, Load (ETL) process to get the latest stats every week.

These repeat users are asking the following when they consider using your API: Can I count on you? Will your API be available for me a month from now? A year from now?

Many public API websites have a dusty, neglected feel. They have blog posts that are a few years old, message boards where posts get no response, or feature road maps that are out of date. If your docs are a ghost town, potential users will go elsewhere.

You can signal trust to these users by ensuring that your feedback mechanism is up to date and any questions are responded to quickly. A current release history and a clear versioning strategy are also good signs of life.

Data science users may be looking for one-time data loads and have less interest in long-term support. They may be asking different questions: Can I trust your data? Do the definitions match the contents? What level of quality is your data? Signal trust to these users by stating information about the quality of your data, and giving specific commitments on update frequency.

Making Great API Docs

When you begin the documentation for a new API, you can start with the core features discussed in this section. These give users what they need to get the job done. As your API matures, continue to improve the documentation and add features that set yours apart from competitors'.

Core Features

All API documentation should include a few core features. Without these basics, it is unlikely that an API will be used by many consumers. They allow a user to understand the value of the API, get the address and any necessary authorization information, and easily test the API. Terms of service and a feedback mechanism ensure that users use the API appropriately and know where to go for help. Following are the core feature that are expected for someone to use the API:

Getting started
> This is an overview or introduction that explains the purpose of the API and how consumers can interact with it. It also provides the address of the API, security requirements, and instructions for requesting user IDs or API keys.

Endpoint definitions
> These explain the purpose of each endpoint, along with the format of requests, responses, and any errors. An endpoint is the full address and resource name that API users will make a call to. For example, *https://api.sportsworldcentral.com/v0/players* is an endpoint that returns a list of players.

OAS file
> The OpenAPI Specification (OAS) file is a machine-readable file that defines the connection instructions and endpoint definitions.

Terms of service

These explain the allowed usage of the API and any rate limits or restrictions that apply.

Feedback mechanism

For people to use your API, they need to have a method to reach out and ask questions. Some common methods include support email addresses, contact forms, and GitHub issue trackers.

SDKs

By creating SDKs in popular programming languages, an API provider can make the process of using the API much easier for consumers. These can help enforce responsible usage of the API, reducing some common problems such as overuse. If an API's primary users are data scientists, this will greatly smooth the process of using your API. You will create an SDK for your API in Chapter 7.

Extra Features

In addition to the basics, several other items improve the DX of the API by simplifying the process of using the API or making it easier to try it out:

Sample program code

These are example snippets of code in popular software languages that demonstrate how to properly use the API.

Interactive documentation

Interactive documentation allows users to submit sample API calls to a development environment to get hands-on experience using the API. Swagger UI is used to generate interactive documentation for FastAPI projects.

Sandbox environment

Going beyond interactive documentation, a full-featured sandbox environment will enable developers to submit multiple API calls in an environment that saves the sample data. For example, they might create an order with one API call and then retrieve the results of the order in another API call.

Additional features

There are a variety of other extras that API providers provide. One that has been increasing in popularity is creating a *Postman collection*, which includes example requests and tests that can be run with the Postman API testing tool.

Some of these require a commitment of time and resources to create, and also to keep current as the APIs change. This is often the role of dedicated developer relations (devrel) staff. API providers must consider the business proposition of the API before investing the resources in these.

> Advanced features can significantly reduce the friction a consumer faces to use the API, and get them up and running quickly. Sometimes this is referred to as reducing the *time to hello world* (TTHW). The quicker a potential consumer can make a basic prototype with your API, the more likely they are to use it.

Reviewing Examples of API Documentation

A tour of some real-world API documentation will demonstrate the features mentioned previously. These examples illustrate some core features and extra features.

Sleeper App

The Sleeper app is a real-life fantasy football league host, like your imaginary Sports-WorldCentral. Its API documentation (*https://docs.sleeper.com*) demonstrates several of the basic features of documentation (Figure 5-1).

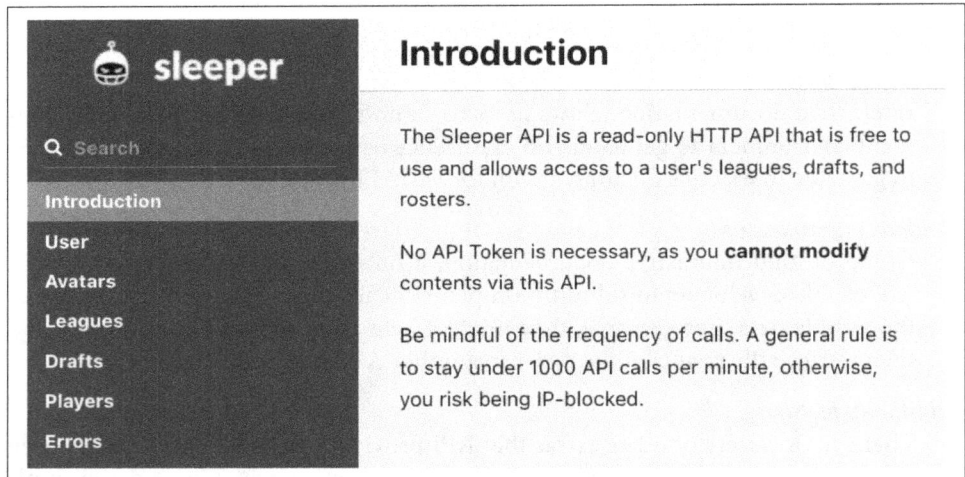

Figure 5-1. Core feature: Getting started page

The landing page gives some basic information about using the API, including authentication requirements (no authentication) and general usage limits (under 1,000 API calls per minute). The documentation provides a search function and navigation bar showing the major endpoints and errors. Figure 5-2 shows the more detailed information available for the Drafts endpoint.

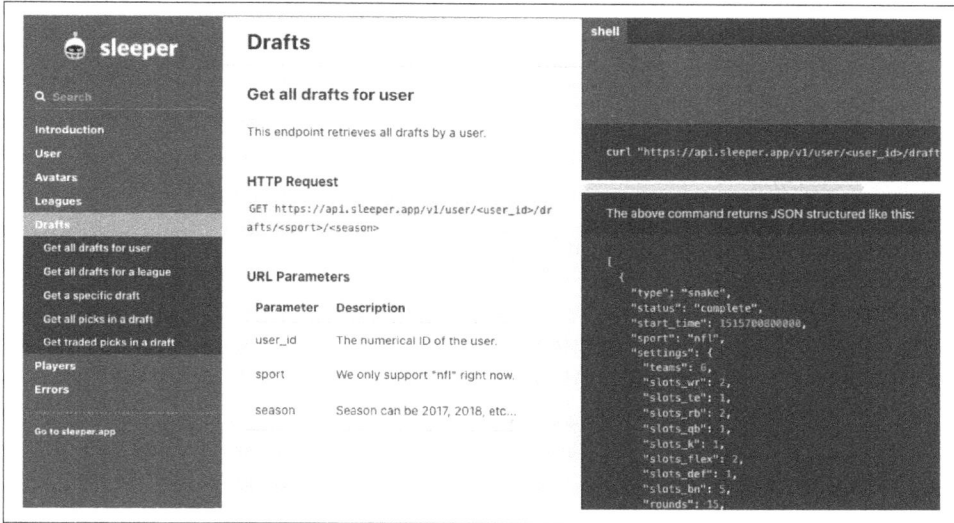

Figure 5-2. Core feature: Endpoint definitions

This page includes the production API URL along with the HTTP verb and required parameters. The righthand column demonstrates an example API call and the output of the API call in JSON format.

MyFantasyLeague

MyFantasyLeague is a fantasy league host that provides a full-featured developer portal (*https://oreil.ly/t7tSf*) and is known for providing quality support to developers.

The welcome page begins with a focus on release notes of API versions (Figure 5-3). It provides links to a request reference page and sample code, and then prominently features the terms of service for using the API.

WELCOME TO THE MYFANTASYLEAGUE.COM DEVELOPERS PROGRAM!

We want to thank you for your interest in the MyFantasyLeague.com Developers Program. The goal of this program is to provide the tools for our community to enhance the overall experience and enjoyment of the entire MFL user base. If you are a developer, these pages should provide you with all the information you need to create applications and other enhancements using the MFL platform.

New! For a list of changes made for the 2022 season, please check the Release Notes (last updated on July 25, 2022). 2020 Release Notes are here and 2019 Release Notes are here

Here is the overall information. You can get details on all the available calls on the Request Reference Page, view sample code or test the requests.

Figure 5-3. MyFantasyLeague welcome page

Following the "test the requests" link from the welcome page brings you to the interactive API documentation (Figure 5-4). What is the difference between interactive API documentation and a sandbox environment? Interactive documentation supports test API calls, but it doesn't store the results between test calls. A sandbox environment keeps track of previous API calls to allow you to test multiple tasks together.

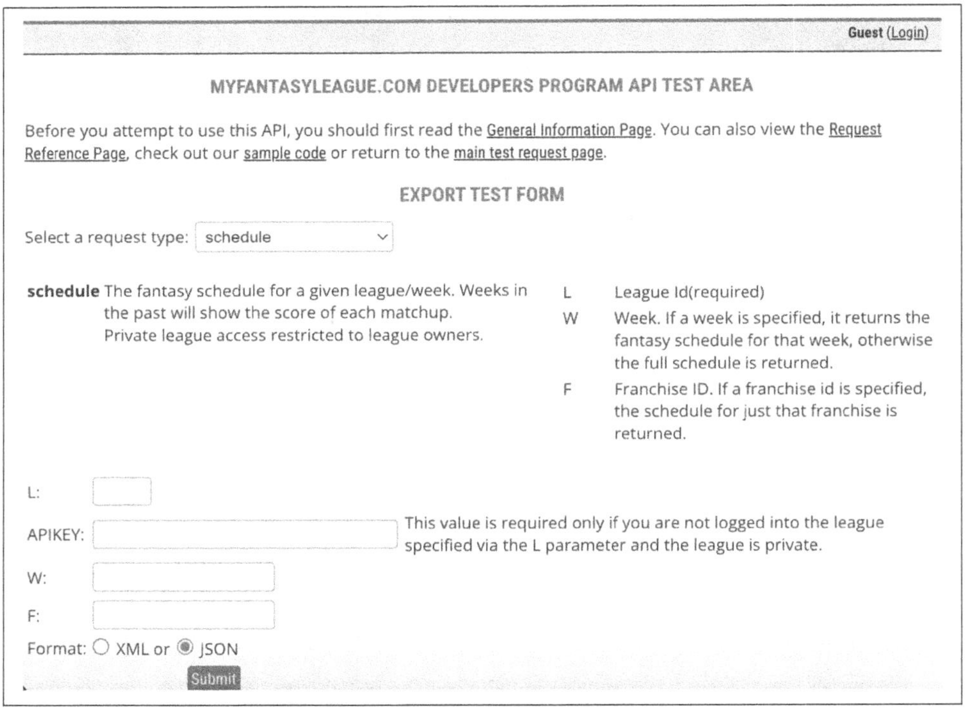

Guest (Login)

MYFANTASYLEAGUE.COM DEVELOPERS PROGRAM API TEST AREA

Before you attempt to use this API, you should first read the General Information Page. You can also view the Request Reference Page, check out our sample code or return to the main test request page.

EXPORT TEST FORM

Select a request type: schedule ⌄

schedule The fantasy schedule for a given league/week. Weeks in the past will show the score of each matchup. Private league access restricted to league owners.

L   League Id(required)
W   Week. If a week is specified, it returns the fantasy schedule for that week, otherwise the full schedule is returned.
F   Franchise ID. If a franchise id is specified, the schedule for just that franchise is returned.

L:

APIKEY: This value is required only if you are not logged into the league specified via the L parameter and the league is private.

W:

F:

Format: ○ XML or ◉ JSON
Submit

Figure 5-4. Extra feature: Interactive documentation

Yahoo! Fantasy Football

Yahoo! is another fantasy football league host that provides detailed API documentation (*https://oreil.ly/4YIxy*). Yahoo! provides sample PHP code for accessing its APIs. This sample code provides detailed instructions on authentication, which can be especially tricky (Figure 5-5).

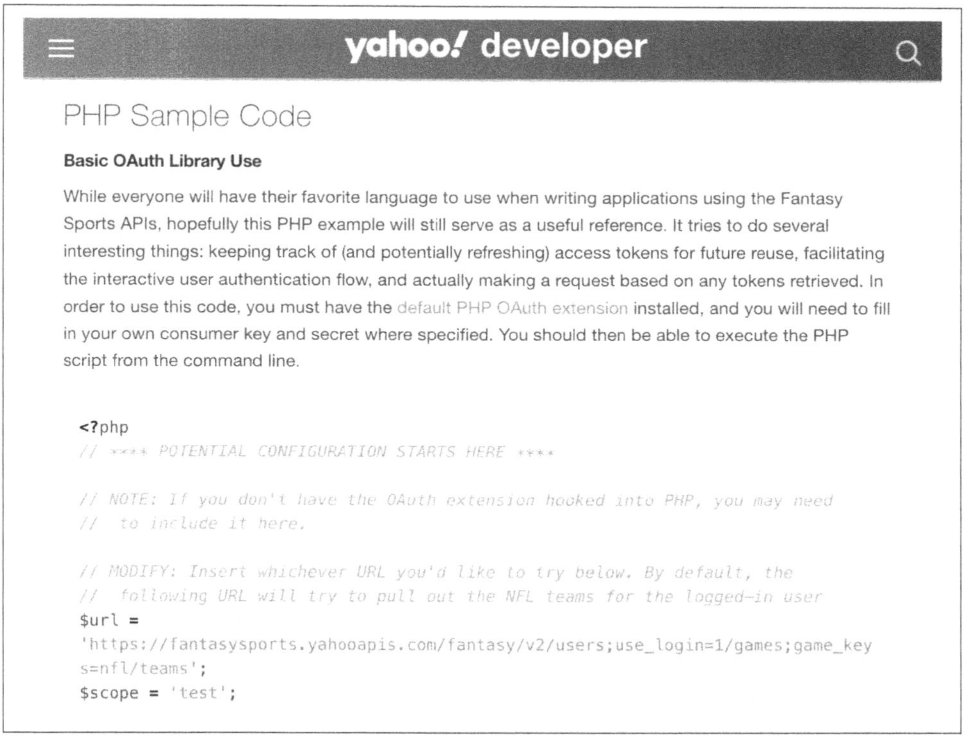

Figure 5-5. Extra feature: Sample program code

Viewing Your API's Built-in Documentation

In Chapter 4, you built an API with the endpoints to fulfill your top-priority use cases. In this chapter, you will view the built-in documentation for the API. You will also make a few updates to the API code to improve the documentation.

Copying Files from Chapter 4

To continue your portfolio project where you left it in the previous chapter, change the directory to *chapter5* and then copy the previous chapter's files into it. The following shows the commands and expected output:

```
.../portfolio-project (main) $ cd chapter5
.../chapter5 (main) $ cp ../chapter4/*.py .
.../chapter5 (main) $ ../chapter4/fantasy_data.db
.../chapter5 (main) $ cp ../chapter4/requirements.txt .
.../chapter5 (main) $ ls *.*
crud.py database.py fantasy_data.db main.py models.py readme.md
requirements.txt schemas.py test_crud.py test_main.py
```

Now launch your API using `fastapi dev` instead of `fastapi run`, as you did in the previous chapter. This tells FastAPI to automatically reload the application each time you make any changes to the program code:

```
.../chapter5 (main) $ fastapi dev main.py
INFO:    Will watch for changes in these directories:
         ['/workspaces/portfolio-project/chapter5']
INFO:    Uvicorn running on http://127.0.0.1:8000 (Press CTRL+C to quit)
INFO:    Started reloader process [9999] using WatchFiles
INFO:    Started server process [9999]
INFO:    Waiting for application startup.
INFO:    Application startup complete.
```

You will see a dialog stating "Your application running on port 8000 is available," as shown in Figure 5-6.

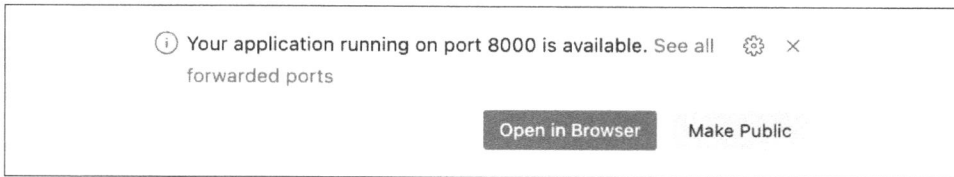

Figure 5-6. Codespaces browser window pop-up

Click "Open in Browser" to open a browser tab outside your Codespaces. This browser will show a base URL that ends in *app.github.dev* that contains the response from your API running on Codespaces. You should see the following health check message in your web browser:

```
{"message":"API health check successful"}
```

Documentation Option 1: Swagger UI

I will go into more detail on the Swagger UI documentation than the Redoc documentation, because the former allows you to test your API. To view the Swagger UI interactive API documentation for your API, copy and paste the following onto the end of the base URL in your browser: **/docs**. For example, the full URL in the browser might be *https://happy-pine-tree-1234-8000.app.github.dev/docs*. You should see documentation as shown in Figure 5-7.

You may recognize this list of endpoints from Table 4-1 in the last chapter. You can see that all the endpoints you implemented in your API are listed here. You can click on any of these endpoints to expand the section that is specific to that endpoint and interact with it.

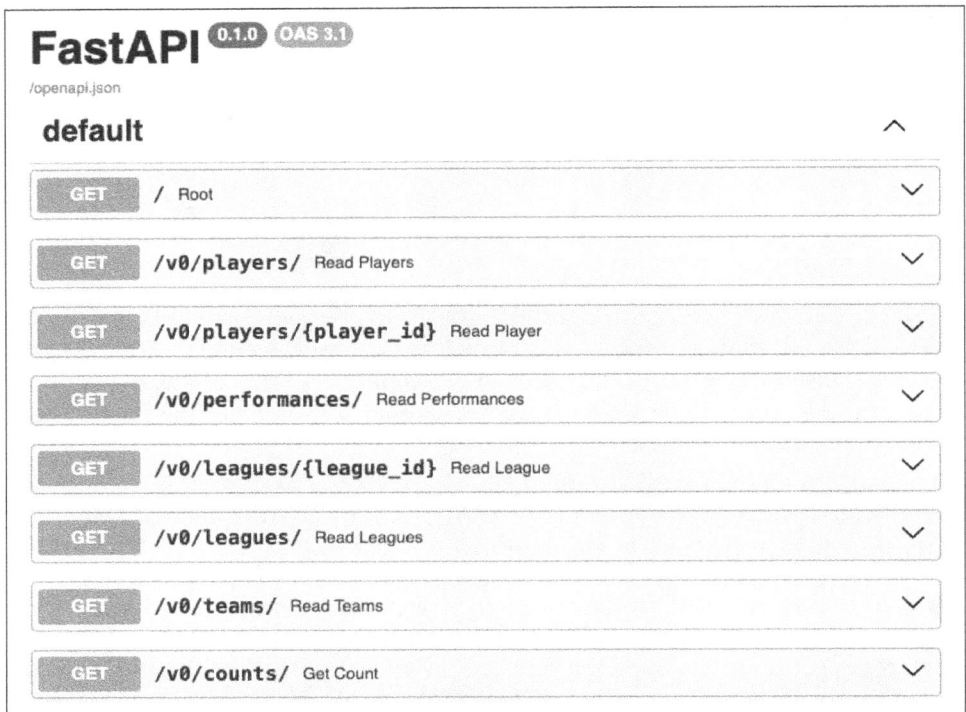

Figure 5-7. Initial Swagger UI interactive documentation

Click on the bar that says "Get /v0/players/{player_id}" and then click the "Try it out" button. This button changes to Cancel, as seen in Figure 5-8.

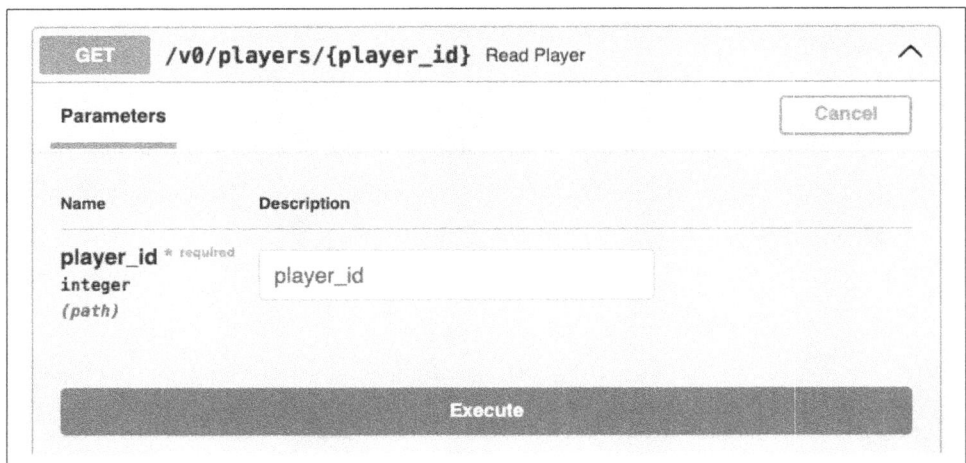

Figure 5-8. Expanded endpoint section, top half

The Parameters section includes the `player_id` parameter that is defined for this endpoint. This parameter is required, it has a data type of integer, and it is a path parameter, which means that when this API is called, the URL path includes the `player_id` value, as shown in the endpoint definition. Figure 5-9 displays the bottom half of this endpoint section.

Figure 5-9. Expanded endpoint section, bottom half

This section describes two expected responses that consumers should plan for. If a request to this endpoint is successful, an HTTP status code 200 (successful response) will be returned, and the response body will be in this format:

```
{
  "player_id": 0,
  "gsis_id": "string",
  "first_name": "string",
  "last_name": "string",
  "position": "string",
  "last_changed_date": "2024-04-27",
```

```
  "performances": []
}
```

However, if the request is invalid, the consumer will receive a response with an HTTP status code 422 (unprocessable content) and an error message in this format:

```
{
  "detail": [
    {
      "loc": [
        "string",
        0
      ],
      "msg": "string",
      "type": "string"
    }
  ]
}
```

To interact with the interactive documentation, click the Execute button. You should see the error result displayed in Figure 5-10, or something similar.

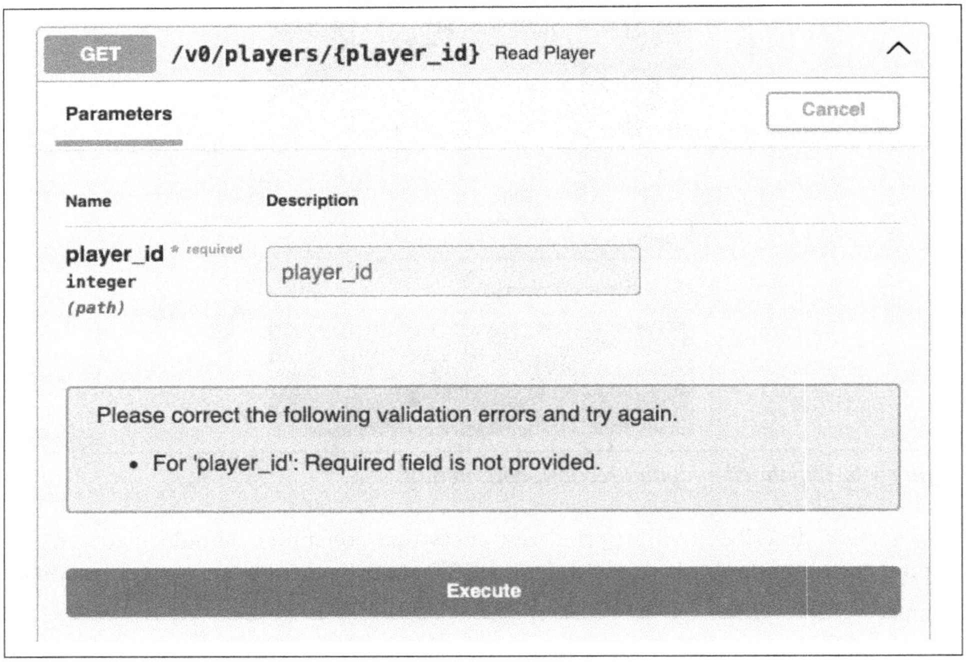

Figure 5-10. Player endpoint error message

The documentation has enforced the required `player_id` parameter and displayed the message "Required field is not provided." Enter a value of **1385** in the `player_id` field, and click Execute again. This request should be successful, and the Responses section of the web page should look similar to Figure 5-11.

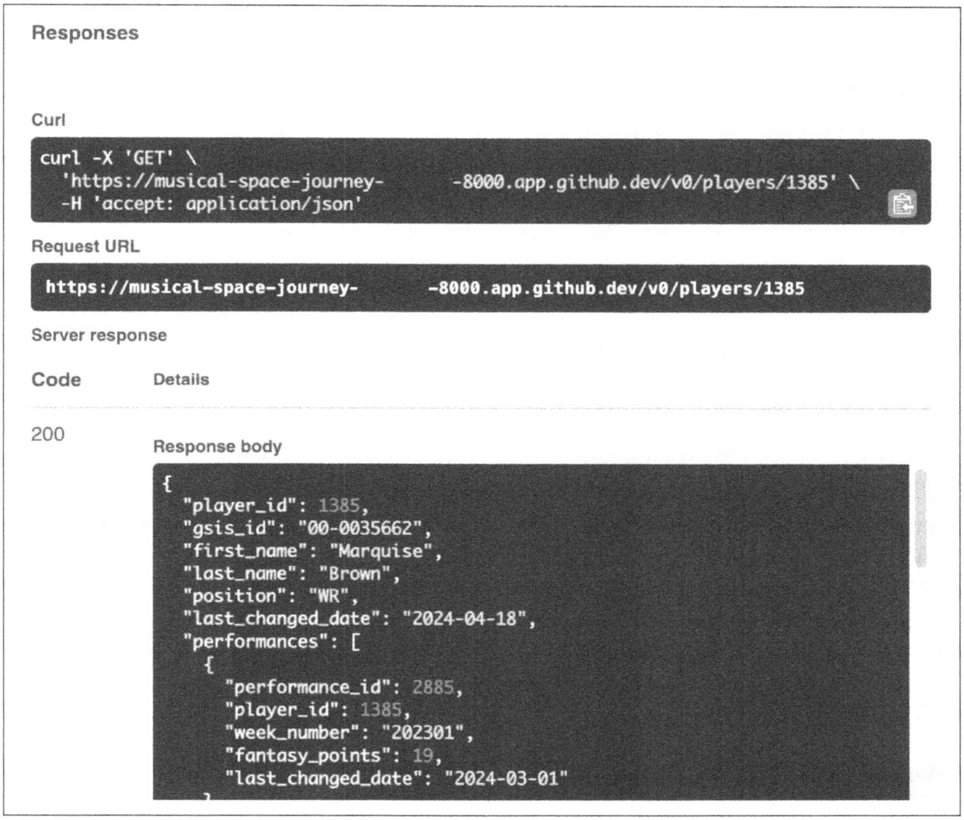

Responses

Curl

```
curl -X 'GET' \
  'https://musical-space-journey-          -8000.app.github.dev/v0/players/1385' \
  -H 'accept: application/json'
```

Request URL

```
https://musical-space-journey-          -8000.app.github.dev/v0/players/1385
```

Server response

Code	Details
200	Response body

```
{
    "player_id": 1385,
    "gsis_id": "00-0035662",
    "first_name": "Marquise",
    "last_name": "Brown",
    "position": "WR",
    "last_changed_date": "2024-04-18",
    "performances": [
        {
            "performance_id": 2885,
            "player_id": 1385,
            "week_number": "202301",
            "fantasy_points": 19,
            "last_changed_date": "2024-03-01"
```

Figure 5-11. Successful player response

This part of the page shows useful information about the HTTP request that is constructed by the documentation and the HTTP response that is sent back by your API.

First, look at the response. In the `curl` section, the documentation displays the command-line statement that could be used to call this API. This is helpful to understand the exact HTTP request that is constructed by the parameters you entered earlier. (cURL is a common command-line utility used to make HTTP requests to web applications and APIs.) You could copy that URL into your browser bar and make the same API call directly.

Remember, the endpoint is a combination of an HTTP verb and a URL. In Figure 5-11, the HTTP verb is GET, as displayed in the documentation. The URL points to the address that Codespaces generates.

In the Request URL section, you see the URL without the HTTP verb. Note that the end of the URL is *v0/players/1385*. This matches the expected */v0/players/{player_id}* from our program code and the earlier documentation. Since player_id is a path parameter, it is added to the end of the URL.

Now let's look at the response. Under "Server response," you see that the response received was an HTTP 200, which is a successful response. The "Response body" shows the JSON data that was returned by the API. It matches the format of the Player object, and it contains the data for player_id 1385, which is what you entered in the path parameter. It also includes the Performance records that are associated with this Player.

The "Response headers" section displays the HTTP headers, which are additional metadata the API sent along with the response body. This is information you would not see if you were calling the API directly in your browser.

As you have seen, this interactive API documentation is very powerful. It compares well to the features of any of the API documentation demonstrated on the real-world fantasy league hosts in this chapter. Since you did not create this code yourself, where is it coming from? This documentation is generated by Swagger UI (*https://oreil.ly/ r6P--*), which is an open source project led by SmartBear software. The Swagger UI has been heavily used in API documentation for many years. Many times when people use the term *API documentation*, Swagger UI is what they mean. It is not an exaggeration to say that this open source project deserves a lot of credit for expanding the popularity of REST APIs as a core technology across the IT industry.

Swagger UI is included in the FastAPI code. It generates this documentation automatically from the OpenAPI Specification (OAS) file that is named *openapi.json* and is linked under the title of this page. FastAPI generates this OAS file automatically from the *main.py* code.

If you look at this endpoint definition in *main.py* that you developed in Chapter 4, you will see the following:

```
@app.get("/v0/players/{player_id}", response_model=schemas.Player)
```

Looking again at Figure 5-8, you can see that the HTTP verb, URL, and path parameter are generated from `@app.get("/v0/players/{player_id}"`.

Looking at Figure 5-9, you can see that the Successful Response definition is generated from the Pydantic schema Player, which you defined in *schemas.py*. The time you spent defining each piece of the Python code continues to pay off in this well-crafted API.

Documentation Option 2: Redoc

In addition to Swagger UI, FastAPI includes a second API documentation option: Redoc. Redoc (*https://oreil.ly/redoc*) is an open source API documentation product that is led by Redocly.

To view the Redoc API documentation for your API, copy and paste the following onto the end of the base URL in your browser: */redoc*. For example, the full URL might be *https://happy-pine-tree-1234-8000.app.github.dev/redoc* in the browser. Click Read Player in the navigation pane on the left. You should see documentation as shown in Figure 5-12.

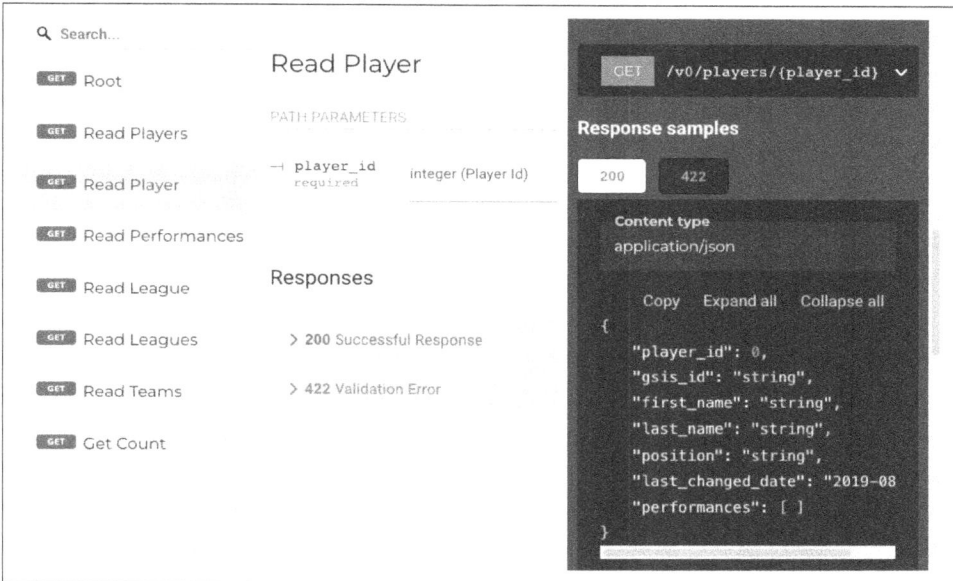

Figure 5-12. Redoc documentation page

As you can see, the Redoc documentation has an attractive three-column layout with the API endpoints in the navigation bar on the left, along with a search function. The center column has the bulk of the information about the endpoint, including parameters, responses, and errors. The righthand column displays example response bodies for successful (200) and error (422) responses.

In general, the contents of the documentation are quite similar to Swagger UI, because both are generated from the same OAS file. The biggest difference in functionality between the two included documentation options is that the Redoc library included in FastAPI doesn't provide a method for submitting API requests. So it is not an interactive option.

What advantages does Redoc have over Swagger UI? In my opinion, the mobile browser layout of Redoc is more attractive than that of Swagger UI. While I generally don't consider mobile functionality important for a technical reference page like API docs, this might be important to some organizations. In addition, the Redoc layout and appearance might be preferred.

Working with Your OpenAPI Specification File

Although these documentation pages are quite impressive, they are nothing without the OAS file. This is literally true, since FastAPI and Pydantic generate the OAS file at the */openapi.json* path, and Redoc and Swagger UI generate their documentation from this file.

But the OAS file is more than simply a way to generate documentation: it is a powerful API definition standard that allows many other tools to interact with your API. One quick example is that in Part III you will create a custom action that enables ChatGPT to answer questions using the data from your API. What information does it need for this? Your OAS file. This is just a drop in the bucket of what the OAS file is used for. The OpenApi.Tools (*https://openapi.tools*) website lists code generators, data validators, SDK generators, mock servers, and a dozen other categories of tools that use the OAS file.

For many years, the software development industry was divided between competing vendor-specific API specifications and response formats. The Swagger specification was made open source in 2015 by Swagger, and in 2017 several companies and groups presented a united front at the API Strategy and Practice Conference promoting OAS as the de facto standard for REST API specifications. (Coincidentally, I was a speaker at that conference and got to witness the events unfold.) At the time, Bill Doerrfield (*https://oreil.ly/m6XNK*) summarized the events by saying that the data format wars are over. You'll read more from Doerrfield in Chapter 13.

The OAS format (*https://oreil.ly/J9v0i*) is quite extensive, so this chapter will focus on the items that are implemented in FastAPI. To view the raw OAS file for your API, copy and paste the following onto the end of the base URL in your browser: */openapi.json*. For example, the full URL might be *https://happy-pine-tree-1234-8000.app.github.dev/openapi.json* in the browser. You should see the raw JSON file, which begins with `{"openapi":"3.1.0","info":{"title":"Fast API","version":"0.1.0"},"paths":`.

The OAS file defines a single API, but the OpenAPI Initiative is working on a new specification called Arazzo (*https://oreil.ly/arazzo*) that will define workflows involving sequences of API calls. One of the potential use cases for this specification is to assist generative AI applications that are based on *large language models* (LLMs). For more information about these applications, read Part III of this book.

To understand the structure of this file, you will need to view the file somewhere that formats JSON in an easier-to-read format. Two convenient approaches to this would be installing a JSON formatter browser extension or opening the file in your IDE. If you save the file and open it in VS Code, you can choose "Format document" from the context menu to see the formatted JSON. Using a browser extension to format JSON has one advantage: as you make changes to the API that modify the JSON file, they can be refreshed and viewed instead of requiring you to download the file repeatedly. I typically install a browser extension for the convenience of viewing API responses. Take a look at the top two levels of the *openapi.json* file hierarchy:

```
{
    "openapi": "3.1.0", ❶
    "info": { ❷
        "title": "FastAPI",
        "version": "0.1.0"
    },
    "paths": { ❸
        "/": {},
        "/v0/players/": {},
        "/v0/players/{player_id}": {},
        "/v0/performances/": {},
        "/v0/leagues/{league_id}": {},
        "/v0/leagues/": {},
        "/v0/teams/": {},
        "/v0/counts/": {}
    },
    "components": { ❹
        "schemas": {
            "Counts": {},
            "HTTPValidationError": {},
            "League": {},
            "Performance": {},
            "Player": {},
            "PlayerBase": {},
            "Team": {},
            "TeamBase": {},
            "ValidationError": {}
        }
    }
}
```

❶ This is a `string` that contains the OpenAPI specification version. For your API, the version is 3.1.0, which was published in 2021.

❷ This field is an `Info` object, which provides metadata about the API. The API's `title` and `version` are the required fields, and they are populated in this example. Optional fields include _summary+, `description`, `termsOfService`, `con tact`, and `license`.

❸ This field contains the list of `paths`, which are relative URLs for API endpoints. You may notice that this list does not contain the HTTP verbs, which make up the other half of an API endpoint. The verbs and additional details are contained one level down inside the `path` objects.

❹ This field can contain a wide range of reusable items that can be referenced in other parts of the OAS. In this example, it contains schemas, which are data structures used by the API.

To see the details of one `path`, expand the */v0/players/{player_id}* item. This `path` object has only one field: `get`, which is an operation object. Expand `get`, and you should see the following:

```
"/v0/players/{player_id}": {    ❶
    "get": {
        "summary": "Read Player",
        "operationId": "read_player_v0_players__player_id__get",    ❷
        "parameters": [
            {
                "name": "player_id",
                "in": "path",
                "required": true,
                "schema": {
                    "type": "integer",
                    "title": "Player Id"
                }
            }
        ],
        "responses": {
            "200": {
                "description": "Successful Response",
                "content": {
                    "application/json": {
                        "schema": {
                            "$ref": "#/components/schemas/Player"    ❸
                        }
                    }
                }
            },
            "422": {
```

```
                    "description": "Validation Error",
                    "content": {
                        "application/json": {
                            "schema": {
                                "$ref": "#/components/schemas/HTTPValidationError"
                            }
                        }
                    }
                }
            }
        }
    },
```

❶ You will recognize that this section contains the information that was displayed in Swagger UI in Figure 5-9. Swagger UI generated it from this section of the OAS file.

❷ This is an operation identifier. This was not displayed in Swagger UI, but some tools use it when processing the OAS. This is also useful to generative AI applications that will process the file.

❸ This is a reference to the `Player` schema that is defined in the `Components` section. These references allow for data structures to be defined a single time in the OAS and referenced in other parts of the file.

Continuing Your Portfolio Project

Although the default information generated in the OAS file is very powerful, you can make it more complete by updating the FastAPI code that generates it.

Table 5-1 summarizes the enhancements you will make to the OAS file by modifying the *main.py* file in your project.

Table 5-1. Updates to the OAS file

Update made	Change in *main.py*	OAS field affected
Add API title, version, description	Add elements to `FastAPI()` constructor	`info`
Add path summaries	Add parameters to path function decorator	`paths`
Add detailed path descriptions	Add parameters to path function decorator	`paths`
Add path tags to group endpoints in Swagger UI	Add parameters to path function decorator	`paths`
Add unique path operation IDs	Update the built-in operation IDs	`paths`
Add description to query parameters	Update the parameters in the FastAPI functions	`parameters`

Adding Details to the OAS info Object

The OAS `info` field contains information about the entire API. The default information in the current OAS file is very generic. You will add content that helps consumers of the API understand its purpose. This involves changes to the `FastAPI()` constructor function.

Open the *main.py* file and make the following updates to the initial application constructor:

```
api_description = """ ❶
This API provides read-only access to info from the SportsWorldCentral
(SWC) Fantasy Football API.
The endpoints are grouped into the following categories:

## Analytics
Get information about the health of the API and counts of leagues, teams,
and players.

## Player
You can get a list of NFL players, or search for an individual player by
player_id.

## Scoring
You can get a list of NFL player performances, including the fantasy points
they scored using SWC league scoring.

## Membership
Get information about all the SWC fantasy football leagues and the teams in them.
"""

#FastAPI constructor with additional details added for OpenAPI Specification
app = FastAPI(
    description=api_description, ❷
    title="Sports World Central (SWC) Fantasy Football API", ❸
    version="0.1" ❹
)
```

Notice the following items that you changed:

❶ The first new statement is the creation of the `api_description` variable with a description of the API. The three quotation marks are used to include a multiline string.

❷ You are passing that `api_description` to the application constructor.

❸ You also give the app a `title`.

❹   And you give the app a version. Notice that you are using a version of 0.1. The major version of 0 communicates that breaking changes can still occur, and it is consistent with the *V0* in the API's URL. I typically start all new API projects with a zero version. The minor version of 1 communicates that this is the first published iteration of this API.

Adding Tags to Categorize Your Paths

Now you move to changes to each individual endpoint, which are described by the path object in the OAS file. The first change to the paths is the addition of *tags*, which are general-purpose attributes that can be used for a variety of purposes. In this case, the reason to add them is that Swagger UI uses them to group API paths together. This is extremely helpful to consumers when an API has several dozen endpoints. To make this change, you will add another element to the path function decorator.

Here is what the changes will look like for the @app.get("/v0/players/{player_id}") decorator in *main.py*. Make these changes to this decorator:

```
@app.get("/v0/players/{player_id}",
        response_model=schemas.Player,
        tags=["player"]) ❶
```

❶   This tag will be used to group endpoints into categories.

You have updated one endpoint. Now use the tag values in Table 5-2 to update the endpoints in *main.py*.

Table 5-2. Tags for endpoints in main.py

Endpoint	tag
/	analytics
/v0/players/	player
/v0/players/{player_id}	player
/v0/performances/	scoring
/v0/leagues/	membership
/v0/leagues/{league_id}	membership
/v0/teams/	membership
/v0/counts/	analytics

Adding More Details to Individual Endpoints

Next, you will add information about individual endpoints. In the OAS file, these are contained in the paths field. To make updates, you will modify each path function decorator. Here are the changes that will be made for each of the paths:

Add a summary
> This summarizes the path and will be displayed on the operation title by Swagger UI.

Add a description
> This gives any additional details about the path and will be displayed below the title by Swagger UI.

Add a description to the 200 response
> This replaces the default "Successful response" with a clearer description. It will be used when the HTTP 200 successful message occurs.

Modify the `operationID`
> This standardizes the `operationID` value, which will be used by a variety of tools that are using the OAS file.

Here is what the changes will look like for the `@app.get("/v0/players/{player_id}"` decorator in *main.py*:

```
@app.get("/v0/players/{player_id}",
        response_model=schemas.Player,
        summary="Get one player using the Player ID, which is internal to SWC",❶
        description="If you have an SWC Player ID of a player from another API
        call such as v0_get_players, you can call this API
        using the player ID", ❷
        response_description="One NFL player", ❸
        operation_id="v0_get_players_by_player_id", ❹
        tags=["player"])
```

❶ Improved summary of the endpoint.

❷ Very detailed description, to help developers and AI use it correctly.

❸ Improved description of the response. Keep this under 300 characters to make ChatGPT happy.

❹ Custom operation ID to replace the auto-generated one.

Apply these changes to the rest of the endpoints in *main.py*.

Adding Parameter Descriptions

Another recommendation for AI to use the API is to add descriptions to parameters used to call the APIs. You will replace the default values of the parameters with the `Query()` statement and a description.

The `read_players` function is beneath the decorator you updated. To add descriptions for the parameters, update that function now:

```
def read_players(skip: int = Query(0, description="The number of items to
skip at the beginning of API call."),
  limit: int = Query(100, description="The number of records to return
  after the skipped records."),
  minimum_last_changed_date: date = Query(None, description="The minimum date of
  change that you want to return records. Exclude any records changed before
  this."),
  first_name: str = Query(None, description="The first name of the players
  to return"),
  last_name: str = Query(None, description="The last name of the players
  to return"),
    db: Session = Depends(get_db)):
```

You have modified the `read_players` function. Update the rest of the API endpoints to add descriptions to all the query parameters.

Viewing the Changes in Swagger UI

With all of these changes made to the OAS file, Swagger UI has much more information than it did previously. Open the */docs* endpoint and you will first notice that the details added to the `info` field are displayed prominently at the top of the documentation, as shown in Figure 5-13.

Figure 5-13. Updated Swagger UI information section

Scrolling down, you will see the effect that the `tags` attributes and the endpoint `summary` values have on Swagger UI, as shown in Figure 5-14. The endpoints are now arranged in categories, and their summaries are displayed to explain their purpose.

Figure 5-14. Tags used to group endpoints

With all of the changes you made in this chapter, *main.py* is too long to show here. You can view the updated *main.py* file in the *chapter5/complete* folder.

Regression-Testing Your API

With all the changes you made to your API code, you should *regression-test* your code, which is running your existing tests to make sure you didn't break anything. Regression testing is helpful for coding changes that you make and updates to software libraries that you use in your code.

In the terminal window, press Ctrl-C to stop the API and display the command prompt. Enter the **pytest test_main.py** command and you should see an output that looks similar to this:

```
$ pytest test_main.py
=================== test session starts ===========================
platform linux -- Python 3.10.13, pytest-8.1.2, pluggy-1.5.0
rootdir: /workspaces/portfolio-project/chapter5
plugins: anyio-4.4.0
collected 11 items

collected 11 items

test_main.py                                          [100%]

=================== 11 passed in 1.01s ============================
```

Updating Your README.md

With the updates you made to your built-in Swagger documentation, you are well on your way to providing a good developer experience. But there are a few API documentation features that you have not provided yet, including getting started, terms of service, and example code. You will update your *README.md* file to fill in the gaps of your documentation.

Update *README.md* with the following contents:

```
# SportsWorldCentral (SWC) Fantasy Football API Documentation

Thanks for using the SportsWorldCentral API. This is your one-stop shop for
accessing data from our fantasy football website, www.sportsworldcentral.com.


## Table of Contents

- [Public API](#public-api)
- [Getting Started](#getting-started)
  - [Analytics](#analytics)
  - [Player](#player)
  - [Scoring](#scoring)
  - [Membership](#membership)
- [Terms of Service](#terms-of-service)
- [Example Code](#example-code)
- [Software Development Kit (SDK)](#software-development-kit-sdk)

## Public API
*Coming Soon*

We'll be deploying our application soon. Check back for the public API address.

## Getting Started
```

Since all of the data is public, the SWC API doesn't require any authentication. All of the the following data is available using GET endpoints that return JSON data.

Analytics

Get information about the health of the API and counts of leagues, teams, and players.

Player
You can get a list of all NFL players, or search for an individual player by player_id.

Scoring

You can get a list of NFL player performances, including the fantasy points they scored using SWC league scoring.

Membership
Get information about all the SWC fantasy football leagues and the teams in them.

Terms of Service

By using the API, you agree to the following terms of service:

- **Usage Limits**: You are allowed up to 2000 requests per day. Exceeding this limit may result in your API key being suspended.
- **No Warranty**: We don't provide any warranty of the API or its operation.

Example Code

Here is some Python example code for accessing the health check endpoint:

```
import httpx

HEALTH_CHECK_ENDPOINT = "/"

with httpx.Client(base_url=self.swc_base_url) as client:
    response = client.get(self.HEALTH_CHECK_ENDPOINT)
    print(response.json())
```

Software Development Kit (SDK)
Coming Soon

Check back for the Python SDK for our API.

Additional Resources

For a comprehensive discussion of the role that API documentation and developer portals play in an API effort, read Chapter 7 of James Higginbotham's *Principles of Web API Design: Delivering Value with APIs and Microservices* (Addison-Wesley, 2021).

For a great short overview of API design concepts, I recommend *Designing Web APIs* by Brenda Jin, Saurabh Sahni, and Amir Shevat (O'Reilly, 2018).

The NordicAPIs.com blog (*https://oreil.ly/pyatT*) publishes articles about developer experience (*https://oreil.ly/YGseq*) and API documentation (*https://oreil.ly/xpcUd*) that have a lot of useful information.

For more information about DX metrics, read "Developer Experience: The Metrics That Matter Most" (*https://oreil.ly/id-00*).

To see examples of individuals publishing tips or tools for consuming undocumented APIs, check out Steven Morse's blog (*https://oreil.ly/stmorse*) and Joey Greco's Leeger app (*https://oreil.ly/leeger*).

For more information about using Swagger UI websites, view the Swagger UI installation instructions (*https://oreil.ly/cSPsX*).

For advice about working with your OAS, read Speakeasy's OpenAPI Guide (*https://oreil.ly/beRNI*).

Extending Your Portfolio Project

You provided several of the core and extra features of API documentation in this chapter. To provide a full-featured developer experience, you can create a standalone developer portal website. Use a template like LaunchAny's LaunchAny Minimum viable portal (MVP) template for APIs (*https://oreil.ly/nhw1K*) to create a developer portal for your API:

- Add a Getting Started page with the steps to quickly help consumers make their first API call. (Remember that TTHW metric!)
- Add a Workflows page to explain how consumers can execute multiple related calls to your API to gather information about leagues, teams, and players.
- Add code samples on a Code Samples page.

If you created an alternate API in previous chapters, follow the steps in this chapter to add additional information to the *openapi.json* file for that API.

Summary

In this chapter, you focused on your consumers by providing API documentation:

- You learned about the built-in API documentation that FastAPI provides using both Swagger UI and Redoc.
- You learned about the OpenAPI specification that FastAPI generates and the value of this specification for users.
- You learned about the components of great API documentation.

In Chapter 6, you will focus on deploying your application to the cloud, where real-world consumers can use it.

Deploying Your API to the Cloud

It sounds a little extreme, but in this day and age, if your work isn't online, it doesn't exist.
—Austin Kleon, *Show Your Work! 10 Ways to Share Your Creativity and Get Discovered* (Workman, 2014)

You have made great progress with your first API. You have selected the most important qualities for your users, developed multiple API endpoints, and created user-friendly documentation. In this chapter, you will publish your API to the cloud, where consumers can access it. This is another chance to share what you have been working on.

Benefits and Responsibilities of Cloud Deployment

The *cloud* is an informal term to refer to the collection of computer servers and connecting infrastructure that make the public internet. For a developer or data scientist, a great way to deploy prototypes and personal projects is through public cloud hosts. Before these became available, the process of running side projects was laborious and the capabilities were pretty limited. As a simple example, before cloud hosts became available to developers, hosting your software on a server required purchasing the physical server. When you finished with your project, you still had that server sitting around. With the cloud you can spin up a virtual server with a few commands, and when you are finished you can delete it. Now you have a great opportunity to deploy your work to a public cloud host with all of the networking and application hosting power that you can imagine (or at least afford).

Benefits

Cloud platforms allow you several great opportunities for the project in this book:

- You can learn the end-to-end process of cloud development, deployment, and operations with the project that you developed.

- You can share it with others as part of your portfolio project. This fast feedback loop allows you to learn and improve much more quickly.

- With the published API, you can use internet-facing tools and products to explore from the user's perspective. This is the focus of Part II of this book.

- You can use generative AI services to consume the API. This is the focus of Part III of this book.

I have found that deploying an application to the web makes the project real in a way that isn't possible when it is only running in a development environment.

Responsibilities

You should be aware that deploying to the internet comes with a few responsibilities as well, the first of which is cost. Cloud providers offer paid services, and they generally charge a variable amount based on the amount of usage of the applications that are hosted. Although some of the cloud providers used in this chapter have a free tier of services or starter credits, they will require a debit or credit card to be added to the account. When you are developing projects on a cloud host, it's not uncommon to have a surprise bill at the end of the month for several dollars. There are also horror stories of people who ended up with unexpected bills costing hundreds or thousands of dollars. I will share some tips about cost management, but you should take care not to sign up for any services that you can't afford. (When deploying the example applications in this chapter, I spent less than $1 USD to host them for several weeks on all the services combined. But your costs could be greater.)

Another responsibility is security. Using the phrase *expose your API endpoint* gives you a hint of this responsibility—when you make your work available to the world, bad actors can access it. If your portfolio project contains fantasy sports data, you won't be putting any personal data at risk. However, if you are careless with the credentials you use to connect to the cloud services, you could expose usernames, passwords, and API keys publicly. This could allow a fraudster to use your accounts to run up large bills mining crypto or creating bot armies to attack websites.

There are several ways to control costs when using cloud hosts:

- Review the costs of services before using them. There are usually several services involved in a coding project. Use cost calculators if available.
- Use services with free tiers and free trial periods.
- Create monthly budgets and set up email notifications to notify you when you are approaching budgeted amounts.
- Shut down or delete resources after use.
- When you have finished working with a cloud host, clean up resources and remove your payment method.
- Keep tight control of your login credentials.
- Use short-lived access keys.
- Only activate the permissions for specific services you are using, and disable them after you are done.
- Do not commit any credentials into source control repositories.
- If any credentials do become exposed, deactivate or delete them. Contact support of the cloud host if necessary.

Choosing a Cloud Host for Your Project

I have used a variety of cloud hosts to deploy prototypes and portfolio projects, and their capabilities and pricing models change over time. When I am deciding what host to use, I first decide what my primary focus is at the time. If I am just focused on getting my app out into the world, I choose the host that is simplest to deploy. If I want to learn about a specific cloud host or some of its underlying technical services, I am more willing to put in the time (and often pay a bit in hosting fees) to learn the deployment process for that host. When I am trying to practice some deployment or operational technique such as continuous integration (CI) or containerization, I select a host that supports that particular method and use it to deploy my app.

This chapter includes instructions for deploying to two cloud hosts, and I would encourage you to try out both of them to see the advantages and disadvantages of each. You will begin by using Render, which is fairly simple to deploy. Then, you will install and configure the Docker containerization tool, which you will use to deploy to Amazon Web Services (AWS).

Table 6-1 displays the new tools or services you will use in this chapter.

Table 6-1. New tools or services used in this chapter

Software or service name	Version	Purpose
Amazon Lightsail	NA	AWS virtual cloud server
AWS CLI	2.15	Command-line interface for AWS services
Docker	24.0	Pack and run your application in a container
Render	NA	Cloud hosting provider

Setting Up Your Project Directory

You Can Start from Here

The instructions in this chapter assume that you completed Chapters 2, 3, 4, and 5. If you're starting your coding in this chapter, you will need to perform a couple of steps to catch up. First, you need to create a GitHub Codespace from the book's GitHub repository. Full instructions are available in "Getting Started with Your GitHub Codespace" on page 22.

To catch up on the coding from previous chapters, you can use the completed set of files in the *chapter5/complete* directory of your Codespace. If you are using these, use the directory *chapter5/complete* in the setup commands that follow.

To continue your portfolio project where you left it in the previous chapter, change the directory to *chapter6* and then copy the previous chapter's files into it. The following shows the commands and expected output:

```
.../portfolio-project (main) $ cd chapter6
.../chapter6 (main) $ cp ../chapter5/*.py .
.../chapter6 (main) $ cp ../chapter5/fantasy_data.db .
.../chapter6 (main) $ cp ../chapter5/requirements.txt .
.../chapter6 (main) $ cp ../chapter5/readme.md .
.../chapter6 (main) $ ls *.*
crud.py  database.py  fantasy_data.db  main.py  models.py  readme.md
requirements.txt  schemas.py  test_crud.py  test_main.py
```

Using GitHub Codespaces as a Cloud Host

For intermittent cloud hosting, GitHub Codespaces may be sufficient. It will only be available while your Codespace is running, but this can be useful for testing with an external tool or sharing it with someone else to review.

When you first run the API in your Codespace, you will see a dialog stating "Your application running on port 8000 is available," as shown in Figure 6-1. Click Make Public.

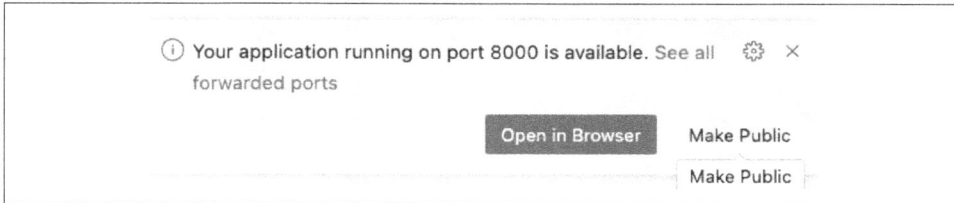

Figure 6-1. Making the API public

The API is now running in Codespaces with a public port. To view the API in the browser, click Ports in the terminal and hover over Port 8000 as shown in Figure 6-2.

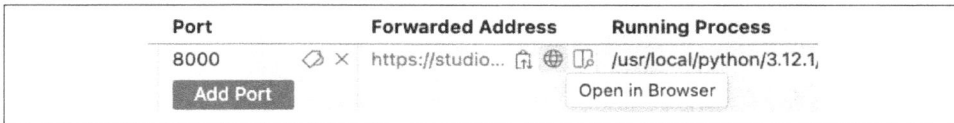

Figure 6-2. Open API on a public address

Click the globe icon, and the web browser should be opened to the health check endpoint of your API. If you look at the address in your browser, you should see a base URL that ends in *app.github.dev*. The browser page should show the response from your API running on Codespaces. You should see the following health check message in your web browser:

```
{"message":"API health check successful"}
```

Your API is running publicly in the cloud.

Deploying to Render

Render calls itself "a unified cloud to build and run all your apps and websites." At the time of this writing, Render has a pricing plan that includes Python hosting for free, except for a small cost for monthly storage. You will be generally following the instructions from Deploy a FastAPI App (*https://oreil.ly/nmdaK*).

The process of deploying to Render only involves a few steps, as shown in Figure 6-3.

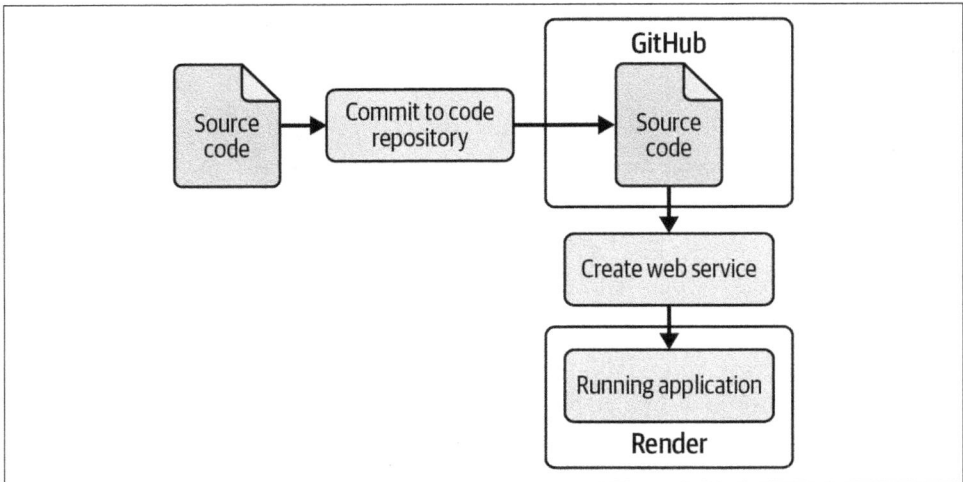

Figure 6-3. Deploying your API to Render

Signing Up for Render

The first step is to sign up for a Render account (*https://oreil.ly/rend*). You can create a new account or use one of the existing services, such as GitHub or Google. It should work the same either way. When you have created your account and provided the information requested, you should see the Render dashboard with no services.

Creating a New Web Service

From the New menu, select Web Service.

On the New Web Service page, select Public Git Repository, enter the URL of your GitHub repo, as shown in Figure 6-4, and click Connect.

Figure 6-4. Choosing how to deploy a web service

The next page should say "You are deploying a Web Service," as shown in Figure 6-5.

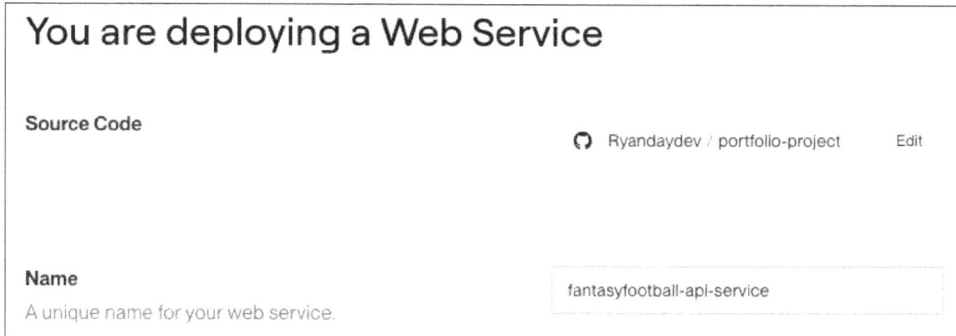

You are deploying a Web Service

Source Code

○ Ryandaydev / portfolio-project   Edit

Name
A unique name for your web service.   fantasyfootball-api-service

Figure 6-5. Render web service settings

Enter the settings on this page as follows:

- *Name*: Enter an available unique name.
- *Project*: Do not create a project.
- *Language*: `Python 3`
- *Branch*: `main`
- *Region*: Select the region nearest you.
- *Root directory*: `chapter6`
- *Build command*: `pip install -r requirements.txt`
- *Start command*: `uvicorn main:app --host 0.0.0.0 --port $PORT`
- *Instance type*: `Free`

Notice that the start command did not use `fastapi run` as you used on the command line. For the web deployments, you are directly using the Uvicorn web server to run the API without using the fastapi-cli library.

Scroll to the bottom of the page, and select Deploy Web Service. A page should be displayed with a deployment log. As you watch, you should see the cloud host installing the software from your *requirements.txt* file and starting an instance of your application. When the process completes successfully, you should see the statement "Your service is live," as shown in Figure 6-6.

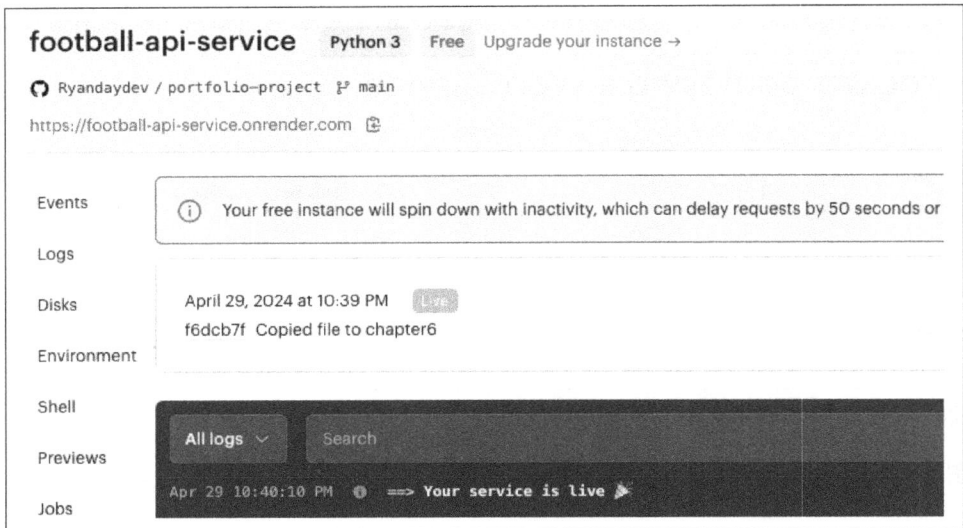

Figure 6-6. Successful deployment on Render

Copy the URL of your web service that is displayed near the top of the window. (In this example, the address is *https://football-api-service.onrender.com*). Paste this URL into another browser window. You should see the health check message of your API. If you access the */docs* endpoint, you can use Swagger UI to check that the rest of the endpoints are returning data. Congratulations! You have deployed your API to the first cloud host!

> If you receive an error in the deployment, verify that you committed the Chapter 6 code to your repository prior to deploying.

Auto-Deploying a Change to Your API

By default, Render is set to auto-deploy any changes you make to files in your repository's *chapter6* folder. Test this by modifying the health-check endpoint in *main.py* with the following message:

```
async def root():
    return {"message": "This is an API health check: status successful"}
```

Commit these changes to your GitHub repository.

Open your web service from the Render dashboard (*https://dashboard.render.com*) and select Events. The most recent event should be a new deployment. After a few minutes, you should see the updated health check text when you access the web service in your browser.

In the next section, you will configure your application to run on Docker, which will be used by AWS Lightsail.

Shipping Your Application in a Docker Container

Whereas Render deployed your application from a source code repository (GitHub), AWS will use an application named Docker. Docker is a very useful tool for shipping applications in *containers*. Just as cargo is shipped in a shipping container, applications are shipped in a software container.

The Docker glossary (*https://oreil.ly/TyStu*) explains a few key terms that you will use. A *Dockerfile* is "a text document that contains all the commands you would normally execute manually to build a Docker image." The *container image*, or Docker image, is "an ordered collection of root filesystem changes and the corresponding execution parameters for use within a container runtime." A *repository* is a set of Docker images. A *container runtime* is software that uses the image to create a container, which is a runtime instance of a container image. You will use Docker as your container runtime.

You will first learn the process of using Docker to run your application locally, and then build on that knowledge to deploy to AWS Lightsail.

Don't worry if all of this information does not make sense yet. As you go through the process of deploying the application in the three different environments, you will begin to see the purpose of the tasks.

Table 6-2 has a summary of Docker commands you will use in this chapter. A full list of commands is available on the Docker cheat sheet (*https://oreil.ly/Prc1P*).

Table 6-2. Docker commands

Command	Purpose
`docker --version`	Verify what version of the library is installed.
`docker build -t`	Build an image from a Dockerfile.
`docker images`	List local images in your environment.
`docker run`	Run a container from a local image.

Verifying Docker Installation

If you are using GitHub Codespaces, Docker should come preloaded for you. Otherwise, follow the instructions on the Get Docker page (*https://oreil.ly/jUGKm*) to install the appropriate version for your development environment.

Verify that Docker is installed by executing the command **docker --version**, which should return the version and build number, such as the following:

```
$ docker --version
$ Docker version 24.0.9-1, build 1234
```

You will perform a couple of fairly simple steps:

- Create a Dockerfile.
- Build a container image from this Dockerfile.
- Run a container based on this image.

Let's get started.

Creating a Dockerfile

A *Dockerfile* contains the instructions that Docker will use to create a container image. Keep in mind that the statements will be executed in the docker build step that you will initiate. Create a file named *chapter6/Dockerfile*:

```
.../chapter6 (main) $ touch Dockerfile
```

Update *chapter6/Dockerfile* with the following contents:

```
# Dockerfile for Chapter 6
# Start with the slim parent image
FROM python:3.10-slim ❶

# set the Docker working directory
WORKDIR /code ❷

# copy from the build context directory to the Docker working directory
COPY requirements.txt /code/ ❸

# Install the Python libraries listed in the requirements file
RUN pip3 install --no-cache-dir --upgrade -r requirements.txt ❹

# Copy the code files and database from the build context directory
COPY *.py /code/ ❺
COPY *.db /code/

# Launch the Uvicorn web server and run the application
CMD ["uvicorn", "main:app", "--host", "0.0.0.0", "--port", "80"] ❻
```

❶ This chooses the *parent image*, which is the official Python 3.10 slim Docker image in this case. This image already contains most of the items you will need to run your application. That allows you to focus on adding your code and any custom libraries.

❷ This statement sets a working directory inside the container image. This working directory is where the remaining commands will be executed unless otherwise specified.

❸ This copies the *requirements.txt* file from *context*, which is a set of files in a location that you will specify when you execute the build statement. (Spoiler alert: in this chapter, you will be using the *chapter6* directory as the context, so the context includes all the files in this folder.)

❹ This uses `pip` to install the specified libraries in your *requirements.txt* file into a new layer in the Docker image. A Docker container image is like a brand-new virtual server: you have complete control of the libraries and versions that are contained in it so that your application works correctly.

❺ These two statements copy your Python program files and SQLite database from the context directory to your Docker working directory.

❻ After steps 1 through 5 have built the container image, this step sets the default command that will be executed each time a container is launched from that image using the `docker run` command. As you did with the Render deployment, you are directly calling Uvicorn to run the API.

These are all the instructions you need to define your container image. The next step will be to build the container image in your local repository.

Creating a .dockerignore File

Just like *.gitignore* excludes files from version control, you want to exclude some files in your directory from the Docker image. Create a file named *.dockerignore* with the following contents:

```
# This is the .dockerignore file for Chapter 6
.gitignore

README.md

*.DS_Store/

**/__pycache__
```

Building a Container Image

To put this Dockerfile to use, enter the following command at the command line:

```
.../chapter6 (main) $ docker build -t apicontainerimage .
```

This tells Docker to build an image (apicontainerimage) using the Dockerfile in the current directory. The container image will be stored in your local Docker repository.

You should see multiple steps being executed. The first time you run this command, this may take several minutes. Future builds will take less time because Docker can cache items that do not change. You should see something similar to the following when it is completed:

```
[+] Building 12.2s (11/11) FINISHED
 => [internal] load build definition from Dockerfile
 => => transferring dockerfile: 670B
 => [internal] load .dockerignore
 => => transferring context: 92B
 => [internal] load metadata for docker.io/library/python:3.10-slim
 => [1/6] FROM docker.io/library/python:3.10-slim@sha256:xxxx
 => [internal] load build context
 => => transferring context: 217B
 => CACHED [2/6] WORKDIR /code
 => [3/6] COPY requirements.txt .
 => [4/6] RUN pip install --no-cache-dir --upgrade -r requirements.txt
 => [5/6] COPY *.py .
 => [6/6] COPY *.db .
 => exporting to image
 => => exporting layers
 => => writing image sha256:xxxx
 => => naming to docker.io/library/apicontainerimage
```

To verify that the image was created successfully, enter the command **docker images** to view the images in your repository. You should see something like the following:

```
.../chapter6 (main) $ docker images
REPOSITORY          TAG       IMAGE ID    CREATED        SIZE
apicontainerimage   latest    x9999       1 minute ago   159MB
```

Running Your Container Image Locally

You use the docker run command to run a container based on this image. There are just a couple of options to notice here. The statement --publish 80:80 maps port 80 inside the Docker container (the second 80) to port 80 on your local environment (the first 80). The statement --name apicontainer1 sets the name of the container that you will be using. This is a convenient way to reference the running container (Docker will also assign an image ID). Finally, apicontainerimage passes the image name that you built in the previous step. Remember, the image is used to run a container.

Execute the following command:

```
.../chapter6 (main) $ docker run --publish 80:80 --name apicontainer1
apicontainerimage+
```

You should see the following:

```
INFO:       Started server process [1]
INFO:       Waiting for application startup.
INFO:       Application startup complete.
INFO:       Uvicorn running on http://0.0.0.0:80 (Press CTRL+C to quit)
```

You will see the message "Your application running on port 80 is available." Select "Open in Browser" to run the API in the web browser. You should see the health check message in the window, exactly like when you ran the API from the command line.

Now if you go back to the terminal window, you will see a log message for the first request to the API:

```
INFO:     1.1.0.1:12345 - "GET / HTTP/1.1" 200 OK
```

Congratulations! Your application is being run by Docker in a container that you defined. To stop the application, press Ctrl-C.

Deploying to AWS

The AWS deployment will take advantage of the Docker container that you created. You will run your application using the Amazon Lightsail service, which is one of the simpler AWS services to get started with.

To begin, create a new AWS account (*https://oreil.ly/09mZ7*) and store your login credentials securely. For this project, it will be acceptable to use the *root user* account, which is the full owner of the account identified by the email address. However, you should enable multifactor authentication (MFA), following the directions in the AWS Identity and Access Management User Guide (*https://oreil.ly/N5Q7e*).

If you continue using AWS, a best practice is to create a separate administration account and use it for normal work instead of the root user.

Creating a Lightsail Container Service

When you have logged in to the AWS console, select Lightsail from the search bar and then select Containers from the lefthand menu. You should see a page that looks like Figure 6-7. Click "Create container service."

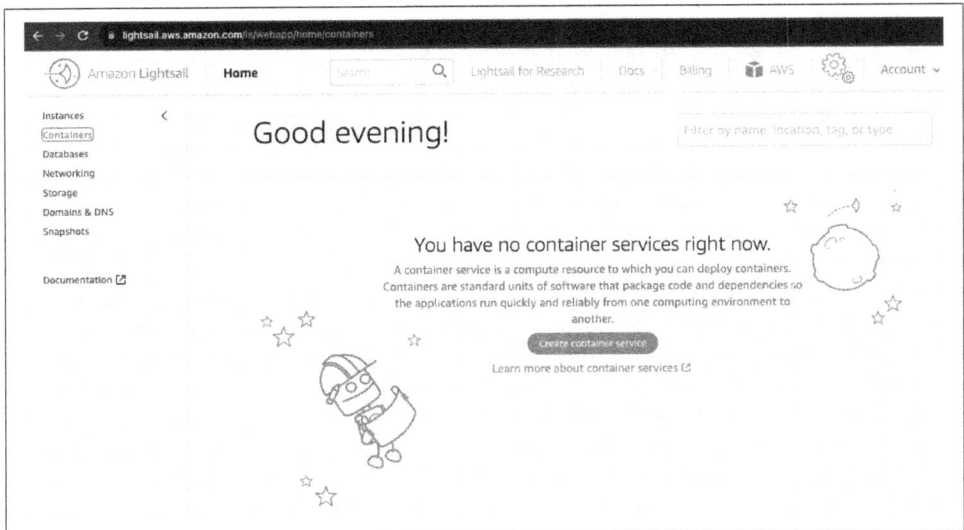

Figure 6-7. Starter page for Lightsail

You will be asked to complete the information for your first Lightsail service:

- *Region*: Select a region that is nearest to you for best performance.
- *Power*: Choose Nano or Micro.
- *Scale*: Choose 1.
- *Identify your service*: Choose a name for your container. This example uses `aws-api-container`.

Click "Create container service". It will take several minutes for the container to be created. When it finishes, you should see a status page like the one in Figure 6-8.

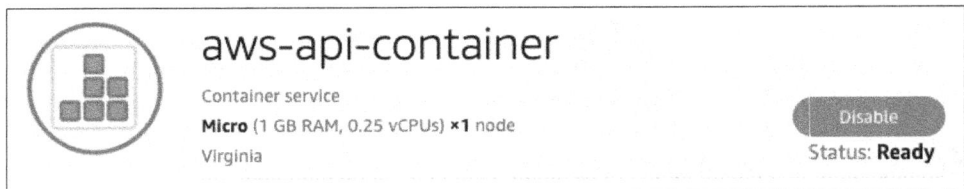

Figure 6-8. Lightsail container service creation completed

Installing the AWS CLI

To interact with AWS, you will use the AWS CLI. Follow the instructions from AWS to install or update the latest version of the AWS CLI (*https://oreil.ly/r8tTp*).

Run the following command to verify the installation:

```
.../chapter6 (main) $ aws --version
aws-cli/2.15.8 Python/3.11.6 Linux/6.2.0-1018-azure exe/x86_64.ubuntu.20
prompt/off
```

Installing the Amazon Lightsail Container Services Plug-in

For Lightsail, an additional installation is required. Follow the instructions on installing the Amazon Lightsail container services plug-in (*https://oreil.ly/xMzkI*).

Configuring Your Login Credentials

To connect the AWS CLI to your AWS account, you will need to configure your authentication and access credentials. There are multiple ways to do this, and it can take a bit of time to set up correctly. As mentioned multiple times in this chapter, you should be extremely careful in the handling of your credentials by cloud services. This includes taking care that they are never committed to a source code repository. Follow the instructions from "Authentication and access credentials for the AWS CLI" (*https://oreil.ly/HAXYD*) to configure your AWS CLI.

Once you have followed the instructions for configuring your credentials, verify them by entering this command:

```
.../chapter6 (main) $ aws sts get-caller-identity
```

You should receive a response in the following format:

```
{
    "UserId": "99999",
    "Account": "999",
    "Arn": "arn:aws:iam::1234:user/username"
}
```

Pushing Your Container Image to Lightsail

Next, you will use the AWS CLI to push a container image to Lightsail. If you run into any issues during these instructions, additional information is available on the "Push, view, and delete container images for a Lightsail container service" page (*https://oreil.ly/lJ0wX*).

To verify that your Docker images are still in your local repository where you built them in the previous sections of this chapter, enter the following:

```
.../chapter6 (main) $ docker images
REPOSITORY          TAG       IMAGE ID       CREATED       SIZE
apicontainerimage   latest    aa0366008bec   2 hours ago   159MB
```

You will push an image from this local Docker repository to Lightsail. Take a moment to understand some of the values you will populate:

Region
> This is the AWS region where you set up the Lightsail service.

ContainerServiceName
> This should be `aws-api-container` if you set up the Lightsail service correctly in the previous steps.

LocalContainerImageName:ImageTag
> This is the name and tag from your local repository. In the preceding example, the value would be `apicontainerimage:latest`.

Enter the following command in the terminal, replacing the values in brackets with your information:

```
lightsail push-container-image --region <Region>
--service-name <ContainerServiceName> --label aws-api --image
<LocalContainerImageName>:<ImageTag>
```

> These commands may wrap in this book, but you will enter them on a single line in the terminal.

Here is an example of the command filled out completely:

```
.../chapter6 (main) $ aws lightsail push-container-image --region us-west-2
--service-name aws-api-container --label aws-api --image apicontainerimage:latest
```

After a couple of minutes, you should see the following:

```
1234: Pushed
12345: Pushed
Digest: sha256:xxxxx
Image "apicontainerimage:latest" registered.
Refer to this image as ":aws-api-container.aws-api.1" in deployments.
```

Go back to the Lightsail console login and view the Images tab on your `aws-api-container`. You should see something like Figure 6-9.

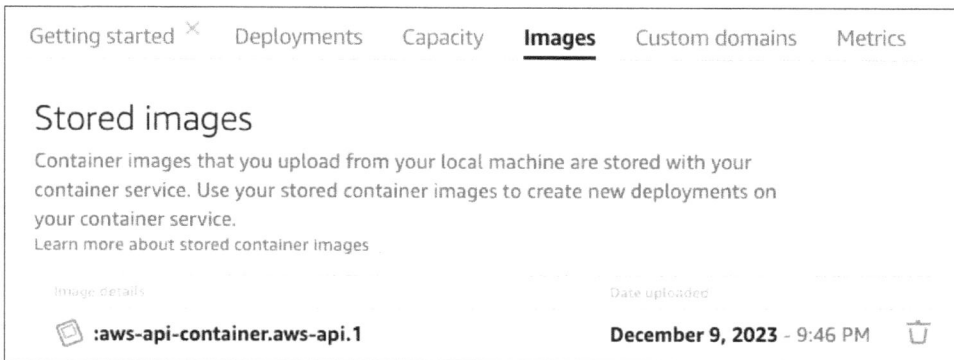

Figure 6-9. Lightsail stored images page

Creating a Lightsail Deployment

On the Deployments tab of the `aws-api-container` service, select "Create your first deployment." Enter the following values, as shown in Figure 6-10:

- *Container name*: Choose a unique container name. This example uses `aws-api-container-1`.
- *Image*: Choose the stored image that was pushed in the previous steps.

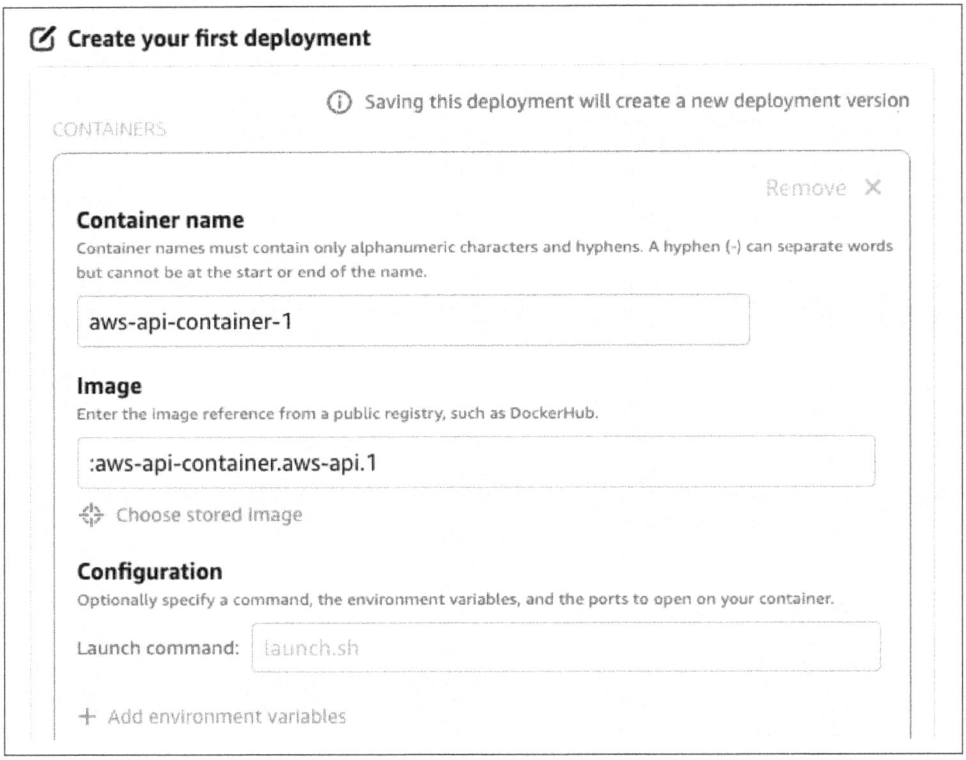

☑ Create your first deployment

(i) Saving this deployment will create a new deployment version

CONTAINERS

Remove ✕

Container name
Container names must contain only alphanumeric characters and hyphens. A hyphen (-) can separate words but cannot be at the start or end of the name.

aws-api-container-1

Image
Enter the image reference from a public registry, such as DockerHub.

:aws-api-container.aws-api.1

⬩ Choose stored image

Configuration
Optionally specify a command, the environment variables, and the ports to open on your container.

Launch command: launch.sh

＋ Add environment variables

Figure 6-10. Deployment options (top of page)

Click "+ Add open ports" and enter a value of 80 in the port. In the Public Endpoint section, click the drop-down box and select the container you created.

After these steps, the completed options should appear as shown in Figure 6-11.

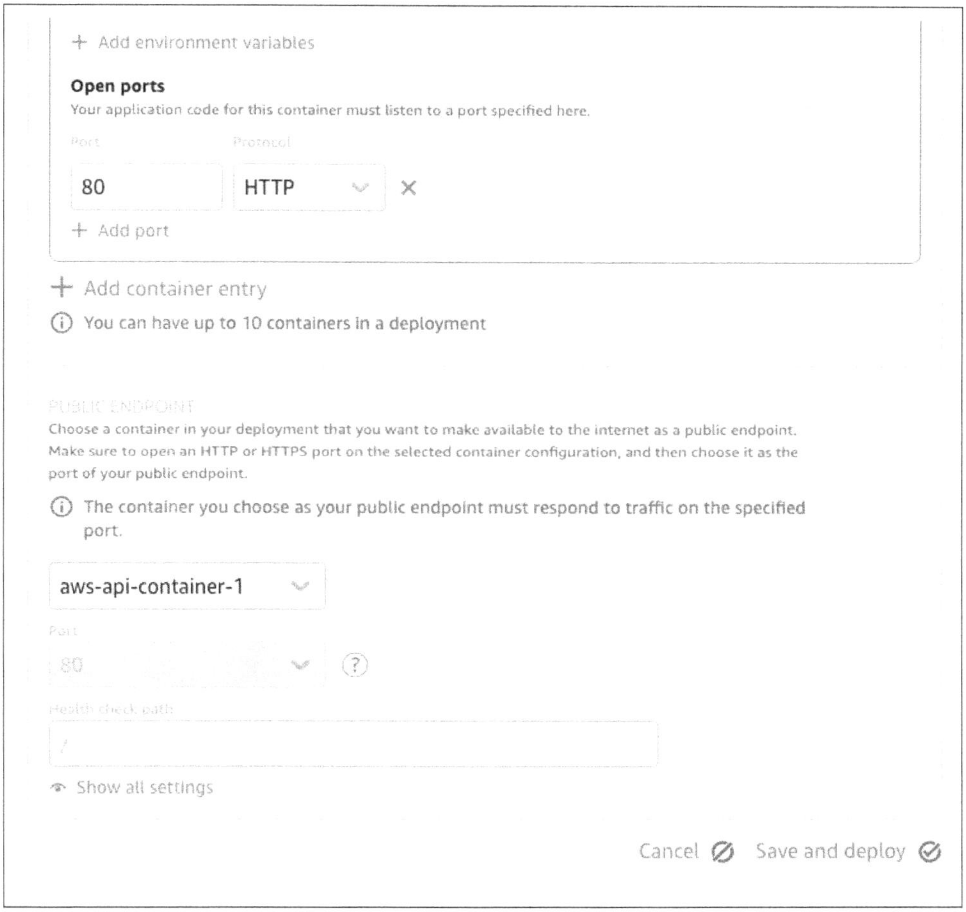

+ Add environment variables

Open ports

Your application code for this container must listen to a port specified here.

Port

Protocol

| 80 | HTTP ∨ | ✕ |

+ Add port

+ Add container entry

ⓘ You can have up to 10 containers in a deployment

PUBLIC ENDPOINT

Choose a container in your deployment that you want to make available to the internet as a public endpoint. Make sure to open an HTTP or HTTPS port on the selected container configuration, and then choose it as the port of your public endpoint.

ⓘ The container you choose as your public endpoint must respond to traffic on the specified port.

| aws-api-container-1 ∨ |

Port

| 80 ∨ | ⑦ |

Health check path

| / |

☞ Show all settings

Cancel ⊘   Save and deploy ⊘

Figure 6-11. Deployment options (bottom of page)

Click "Save and deploy." Lightsail will begin deploying. After a few minutes, you should see a status of Active, as shown in Figure 6-12.

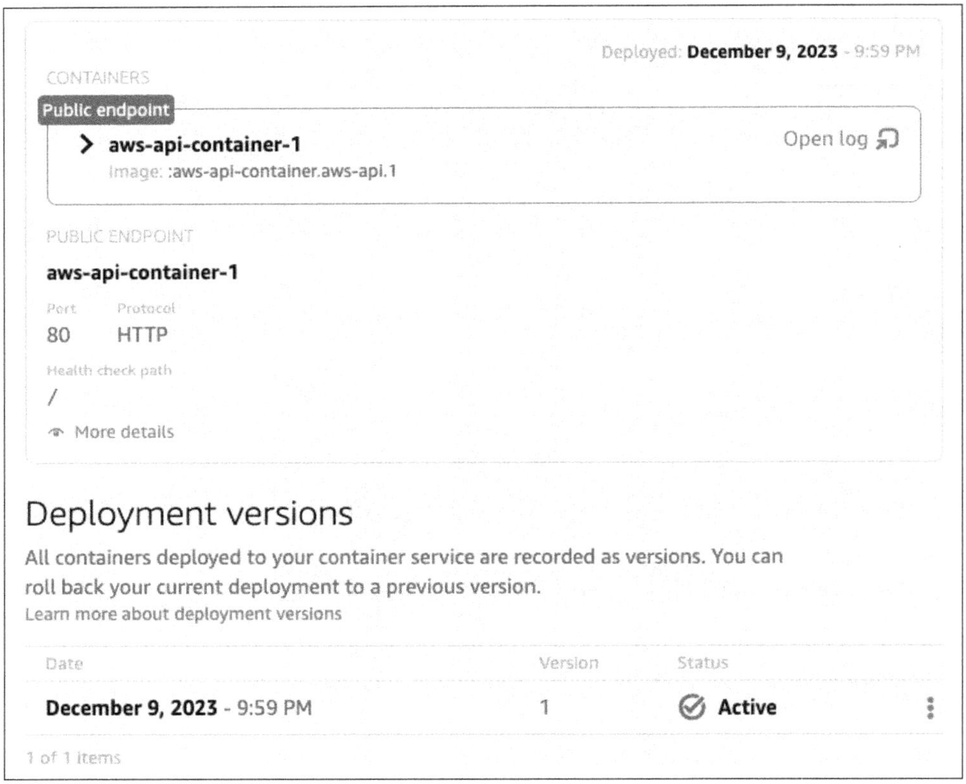

Figure 6-12. Lightsail deployment success

Your container will have a Status of Running, and a "Public domain" will be generated (not shown in Figure 6-12). This is the base URL for the API. To verify that your API is running, copy the value from the "Public domain" and paste it into a browser bar. You should see your API's health check page. Congratulations, you've deployed your API using AWS Lightsail!

Updating Your API Documentation

In Chapter 5, you created the *README.md* file as API's documentation, but you didn't have an API address yet. Update the Public API Address section of the *README.md* file as follows, replacing [`API URL`] with your deployed application's address, wherever you find it:

```
## Public API

Our API is hosted at [API URL]/([API URL]/).

You can access the interactive documentation at [[API URL]/docs]([API URL]/docs).

You can view the OpenAPI Specification (OAS) file at
[[API URL]/openapi.json]([API URL]/openapi.json).
```

Extending Your Portfolio Project

Here are a few ways to use what you've learned in this chapter to extend your portfolio project:

- If you have been creating additional APIs in previous chapters, select one of the cloud hosts from this chapter and deploy those APIs.

- Identify another cloud host from the "Deploy FastAPI on Cloud Providers" (*https://oreil.ly/PsEGM*) page, and deploy your API to one of them.

- After deploying your project to one of the cloud hosts, research and configure a deployment pipeline. Each uses different techniques, but they generally involve creating tasks to redeploy your application each time a change is made to the source code repository.

Additional Resources

The Docker website (*https://oreil.ly/6AyzJ*) provides a deep dive into the concepts around Docker containers.

For best practices on containerizing Python, take a look at "Best practices for containerizing Python applications with Docker" from Snyk.io (*https://oreil.ly/1SYWr*).

Summary

In this chapter, you learned the benefits and responsibilities of using cloud hosting platforms to deploy your API. You now have experience in deploying your application in several different ways:

- In a local development environment or GitHub Codespaces environment
- In a Docker container in a local development environment or GitHub Codespaces environment
- On the Render cloud host
- On AWS using Amazon Lightsail and Docker

In Chapter 7, you will create an SDK to make your API easier to use for Python developers.

Batteries Included: Creating a Python SDK

Make the right things easy and the wrong things hard.
—Kathy Sierra

To create an API data scientists will love, you should give them a *software development kit* (SDK) to call the API with. This is an extra step that most API producers do not take, but it makes life much easier for your users. In this chapter, you will learn the value of SDKs and benefit from practical tips from several experts, and then you will create a Python SDK for the SWC Fantasy Football API. Building an SDK is the capstone of your Part I portfolio project.

SDKs can include code examples, debuggers, and documentation, but the term commonly refers to a custom software library that acts as a wrapper for your API. It allows developers and data scientists to interact with your API directly in their programming language, without requiring extra code to handle API communication.

Figure 7-1 shows how your users will employ an SDK to call your API, instead of calling it directly.

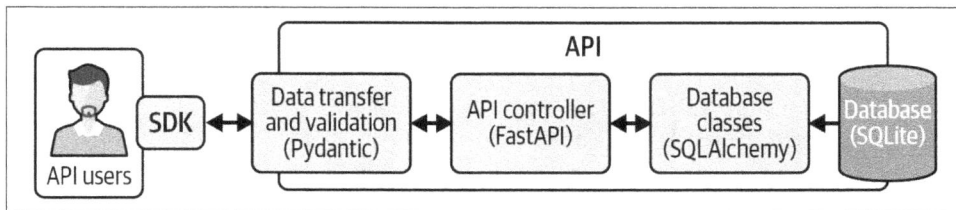

Figure 7-1. SDK interacting with the API

SDK Perspectives: Zan Markan on SDK Fundamentals

Zan Markan is an experienced software engineer and developer advocate from the United Kingdom. Zan has developed SDKs and other tools in his developer relations role. I talked to Zan about SDKs and followed up with additional questions.

What exactly is an SDK?

An SDK is a programming language–native means of exposing an API that would otherwise only be accessible as a REST endpoint. They usually bundle the common REST API calls in functions or classes that you can use directly in code by importing a library.

Can you share an example?

Let's say you have your Premier League SDK. You just want to import your Premier League SDK. There would be a way to add an API key so that every subsequent call would be authenticated easily. Then, the developer can just use the SDK to get a list of fixtures for a day. Or get the scoreboard with a single function call. It takes a lot of complexity away to make building something faster.

How common is it for companies to provide an SDK?

It's more common to provide SDKs if the core audience of your product is someone who codes, such as a developer or data scientist.

SDKs Bridge the Gap

To learn the benefits of SDKs and tips for implementing them, I spoke to several experienced SDK developers and followed up with written questions.

Joey Greco is a software engineer who has created open source SDKs for several fantasy football league hosts, including Sleeper (*https://oreil.ly/8dXo_*) and MyFantasy-League (*https://oreil.ly/SIPsV*). He explains how SDKs help users: "A well-built SDK takes care of all the nitty-gritty for you," he said. "A well-built SDK gives me a few lines of code I can copy/paste on my machine along with a few examples of how to access and manipulate various data. It tells me what I need to do to authenticate (pass your API token into this function, etc.). It's a great way to bridge the gap from an external service to the code that you're writing."

It's useful to remember that developers and data scientists aren't using your API out of an interest in APIs—they have a job to do, and the API helps do the job. Wrapping the API in prebuilt program code makes that even simpler. I once conducted a usability session with a data-focused user of an API developer portal. I was surprised (and maybe slightly insulted) when she told me she didn't care about my APIs—she just wanted the data. For a user like that who just wants the data, an SDK will be a time-saver.

Also, consider that your users may use multiple APIs and data sources from different providers. The time they spend learning how each API works and configuring their code for it is a diversion from their end goal. And not every developer or data scientist has experience in using APIs in a resilient and secure manner. A mature SDK can provide good software development practices to these users with a minimum of trouble.

Simon Yu, cofounder of the SDK generation platform Speakeasy, shares additional benefits: "All the boilerplate code that API consumers needed to write before is already taken care of by the SDK library itself. Instead of every consumer reinventing the wheel themselves (frustrating, time-consuming, error-prone), they simply import the SDK, call the correct method, and go."

There's another benefit too. When developers use an SDK in their IDE, they will get auto-complete and type hints while they code. This makes their development quicker and also enables generative AI tools such as GitHub CoPilot or AWS CodeWhisperer to generate accurate code for them.

Producers benefit from anything that makes it easier to use their APIs. Like good API docs, SDKs reduce the friction for new consumers, which reduces the time-to-hello-world metric discussed in Chapter 5. Simon Yu said SDKs can be a profitable investment for API producers: "An API consumer (who might also be another large enterprise!) often won't want to pay for a service until they are up and running in production. For many API providers, therefore, unblocking API consumers also unblocks revenue."

Yu said SDKs can also reduce API support expenses: "Since SDKs provide a ready-made way to integrate with the API and eliminate the need to write custom integration code from scratch, they dramatically reduce the support burden required of API producers." He added, "Without SDKs, if an integration doesn't work, API producers often get pulled into 1:1 support, which is extremely costly."

An SDK is a good way to encourage users to use your APIs responsibly by conforming to call limits and sending correctly formatted requests. SDKs make doing the right thing the easy thing.

SDKs can be part of an overall API product strategy. One of the best examples I have seen is by StatsBomb (now Hudl StatsBomb) (*https://statsbomb.com*), a sports data and analytics provider. StatsBomb provides football and soccer data with a strong emphasis on supporting research and education in the sports field. StatsBomb hosts a sports analytics platform (*https://oreil.ly/Orp-g*) that provides in-depth analysis and visualizations about players and teams from around the world. StatsBomb also provides data through a subscription-based API (*https://oreil.ly/CzhyS*) that allows its customers to pull data into their own analytics software. To support researchers and students, it also provides some of its data for free download on its open data GitHub repository (*https://oreil.ly/LPGZM*).

For StatsBomb, SDKs tie the paid and free services together. The company maintains two SDKs for its APIs: statsbombr (*https://oreil.ly/IQO_f*) and statsbombpy (*https://oreil.ly/0VOdS*). As shown in Figure 7-2, paid subscribers can use the SDK to get live data, while nonsubscribers can use the SDK to access static data for free. Internally, the SDK gets the paid data from an API and the free data from a file download. But this complexity is shielded from the SDK users: they just know they get the data they want.

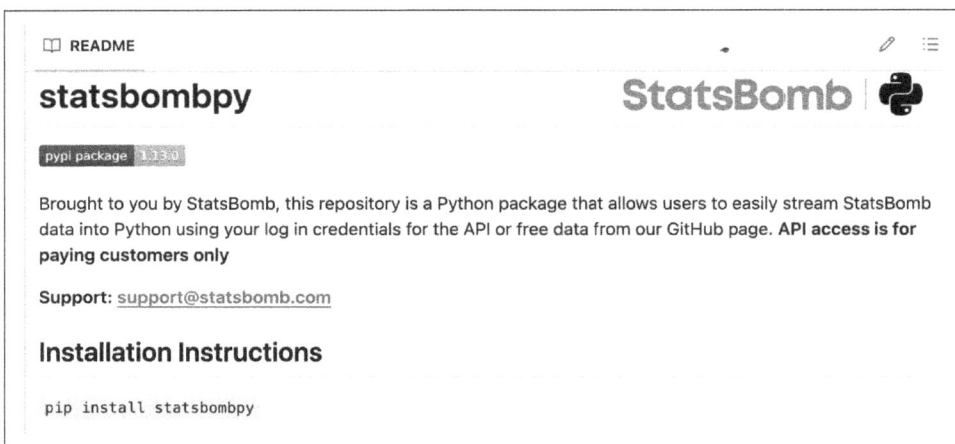

Figure 7-2. The statsbombpy SDK documentation

SDK Perspectives: Francisco Goitia on Implementing SDKs

I talked to Francisco Goitia, lead machine learning engineer at StatsBomb and the developer of the statsbombpy SDK.

What motivated StatsBomb to create SDKs for your APIs?

Our philosophy is to understand the end user of our products. Our user might be a football analyst who doesn't know how to query APIs. They just want to do a plot of the data to get their job done. So, if they can install a Python library with `pip install` and start using the data, it makes their life easier.

For the users of your APIs, why do you think SDKs make their life easier?

If I was writing client code to use our APIs, I would need to write something like the SDK to decouple the API from my software. That's using good software practices. So, by using the SDK, they don't have to do that extra work.

Why did you decide to use R and Python for your SDKs?

Those are the languages that the football analyst community uses, with Python becoming more common recently. When I started in 2020, we already had statsbombr. In the first week I was here, I started creating statsbombpy and our customers started using it.

Picking a Language for Your SDK

Your SDK journey starts by knowing your developer audience and how they will utilize your APIs.

   —SDKs.io

The first decision that you will make when developing an SDK is which programming languages you will support. The SDK does not need to be written in the same language that the API is developed in. Instead, it needs to be written in the language that the API's consumers use. They will import the SDK in their code, and use it to interact with your API, instead of calling the API directly.

API producers sometimes provide SDKs in multiple languages for different consumers. Zan Markan summarized this approach as "Go where the users are." For data scientists, he said, "it's going to be Python, and there might be some R. Python is so heavily used in that community that you want to focus on that." His view is consistent with the "State of Data Science" survey discussed in Chapter 1, which listed Python as the most common language used by data scientists. To get started with a Python SDK, an open source tool like the OpenAPI Python client (*https://oreil.ly/cnR-u*) could be used to generate stubs that you could add more functionality to.

The more languages you create SDKs in, the more maintenance they require. Companies like Speakeasy (*https://oreil.ly/speasy*), APIMatic (*https://oreil.ly/bpMKs*) (publisher of SDKs.io), and Fern (*https://oreil.ly/fern*) provide commercial tools that generate SDKs in multiple languages directly from the OpenAPI Specification (OAS) and keep them updated over time.

Simon Yu explained the benefits of auto-generation services: "Getting the SDK design and implementation right is difficult. Maintaining and supporting them is even harder. What happens if the team member responsible for updating the SDK leaves? Now multiply this problem across all the languages you want to offer the SDK in," he said. "This is where Speakeasy and other SDK generators come in."

> ### You Can Start from Here
>
> The SDK code for your project is separate from the API that you created in earlier chapters. You can start here even if you have not completed earlier chapters in this book. Later in this chapter, you will execute the API code so that you can verify the SDK works with your API. Even if you have not completed the earlier chapters, the working API code is available in the *chapter6/complete* folder.

Starting with a Minimum Viable SDK

The last piece of your Part I portfolio project is to create a Python SDK. Throughout this chapter, you will learn tips and tricks from API experts and implement Python features inspired by them: Francisco Goitia's statsbompy (*https://oreil.ly/LUiJd*), SDKs from Simon Yu's company Speakeasy, and Joey Greco's pymfl (*https://oreil.ly/ut2nK*), along with additional reference documentation.

You will start with a very simple working SDK to make sure the project packaging is working. I'll call this your *minimum viable SDK*. After you verify the package works and can be installed, you will add additional features.

Expert Tip: Making Your SDK Easy to Install

> *...if they can install a Python library with pip install and start using the data, it makes their life easier.*
>
> —Francisco Goitia, StatsBomb

Many programming languages have a standard method of downloading and installing libraries from an external repository, such as npm for Node.js and Maven for Java. Python SDKs are commonly published on the Python Package Index (PyPI) (*https://pypi.org*). Hosting an SDK on PyPI enables Python developers to install the SDK into their environment using the `pip3` tool, as you have done in earlier chapters with libraries such as FastAPI and SQLAlchemy. The `pip3` tool can also install packages directly from a GitHub repository, if the project is structured correctly. This is how you will structure your SDK project and give instructions to users. As a bonus, this structure is what is needed to publish to PyPI if you choose to.

Change the directory to *chapter7* and create the *sdk* folder. Then, change to the *sdk* directory and create a file named *pyproject.toml*:

```
.../portfolio-project (main) $ cd chapter7
.../chapter7 (main) $ mkdir sdk
.../chapter7 (main) $ cd sdk
.../sdk (main) $ touch pyproject.toml
```

The *pyproject.toml* file provides all the settings that tools like `pip3` need to package your code and install it correctly in Python environments. Add all of the following text to your *pyproject.toml* file:

```
[build-system] ❶
requires = ["setuptools>=61.0"]
build-backend = "setuptools.build_meta"

[project] ❷
name = "swcpy"
version = "0.0.1" ❸
authors = [
  { name="[enter your name]" },
]
description = "Example SDK from Hands-On API Book for AI and Data Science"
requires-python = ">=3.10" ❹
classifiers = [
    "Development Status :: 3 - Alpha", ❺
    "Programming Language :: Python :: 3",
    "License :: OSI Approved :: MIT License",
    "Operating System :: OS Independent",
]
dependencies = [ ❻
        'pytest>=8.1',
        'httpx>=0.27.0',
]
```

❶ This section selects setuptools (*https://setuptools.pypa.io/en/latest/setuptools.html*) as the build backend for your project. The Python Packaging Guide lists several other choices for the build backend, including Hatchling, Flit, and PDM. I have chosen setuptools because all of the example SDKs referenced in this chapter use it, and it is simple to use.

❷ This section gives the basics about the package, including name, version, description, and author. (You should put your name there—it's your portfolio project, after all.)

❸ The version should be updated each time major changes are applied.

❹ This communicates the minimum Python version your package supports.

❺ The development status `Alpha` shows that this SDK is in the early stages of development.

❻ This section lists the Python libraries that are required for your package to work. This is the only section in this chapter that will list dependencies—there is no *requirements.txt* file used for the SDK.

Expert Tip: Making the SDK Consistent and Idiomatic

Simon Yu said, "The SDK should be consistent and predictable. Naming conventions, error handling, and response format should be the same throughout the SDK to avoid unnecessary confusion to users." For example, your SDK functions returning a single item will begin with `get_` and functions returning lists will begin with `list_`.

Sdks.io says you should make your SDK *idiomatic*, following the norms used by other programmers of that language. For a Python SDK, this means that your code should be *Pythonic*. This is a broad term, but it means that Python code should follow the conventions that other Python programmers and tools use. Python is a living language, with new features being added all the time, so the coding conventions try to keep up. Conventions are established in the Python community through Python Enhancement Proposals or PEPs. If you'd like to understand the process, check out PEP 1 – PEP Purpose and Guidelines (*https://oreil.ly/dgQ79*).

For the overall style of your SDK, you will be using PEP 8 – Style Guide for Python Code (*https://oreil.ly/oersV*). The official Python docs (*https://oreil.ly/csNt5*) provide a good summary of PEP 8 style: use 4-space indentations, keep lines to 79 characters or less, and use `UpperCamelCase` for classes and `lowercase_with_underscores` for functions and methods.

A few additional Pythonic conventions you will use in your SDK include PEP 202 – List Comprehensions (*https://oreil.ly/7rivi*), PEP 343 – Context Managers (*https://oreil.ly/bgVdx*), PEP 257 – Docstrings (*https://oreil.ly/q5zA2*), PEP 518 – Build System Requirements (*https://oreil.ly/WLYFx*), and PEP 484 – Type Hints (*https://oreil.ly/RkeOY*). These items will be explained as they are added to the code in your SDK.

The *swc_client.py* file is the primary client that will interact with your API. You will start with a very basic client. Create the directory structure, the package initialization file, and the starter Python client:

```
.../sdk (main) $ mkdir src
.../sdk (main) $ mkdir src/swcpy
.../sdk (main) $ touch src/swcpy/__init__.py ❶
.../sdk (main) $ touch src/swcpy/swc_client.py
```

❶ Because your SDK will be a Python package, each directory that contains code will contain a file named *__init__.py*. An empty file can do this job for now.

To begin creating the minimal SDK, give the *swc_client.py* file just enough functionality to call the health check of the SWC API:

```
import httpx ❶

class SWCClient:

    def __init__(self, swc_base_url: str): ❷
```

```
            self.swc_base_url = swc_base_url  ❸

        def get_health_check(self):  ❹
            # make the API call
            with httpx.Client(base_url=self.swc_base_url) as client:  ❺
                return client.get("/")
```

❶ For now, your `import` statement only contains `httpx`, which is the core Python library you will use to interact with the API. The rest of this file defines the `SWCClient` class. Users of the SDK will create an instance of the class when they call the SDK, by executing the `SWCClient()` method.

❷ This is the class constructor, which is executed one time when a client instance is created. The method can be used to initiate methods and variables that are unique to an individual instance of the class, in contrast to the constant variable defined earlier. The `self` parameter is passed as the first parameter to every class method in Python, including the constructor.

❸ The constructor also receives the string parameter `swc_base_url`, which is the base address of the API to call. You set the class variable `self.swc_base_url` so that this value can be used in the `get_health_check()` method.

❹ This function calls the API's health check endpoint.

❺ The `httpx.Client` has quite a few features available, but for now you are setting the base URL and using it to call the "`/`" endpoint, which is your API's health check.

To ensure that you have configured your SDK package correctly, it's time to install it in your Codespace with `pip3`. This will use the new source code and configuration files you created and install it as a library that you can access in Python files. It will also install the dependencies from the *pyproject.toml* file's `dependencies` section.

Now install your package locally with the `-e` option, which ensures that the package is updated locally as your code changes:

```
.../sdk (main) $ pip3 install -e .
Processing /workspaces/portfolio-project/chapter7/sdk
  Installing build dependencies ... done
  Getting requirements to build wheel ... done
  Preparing metadata (pyproject.toml) ... done
...
Successfully built swcpy
Installing collected packages: pluggy, iniconfig, pytest, swcpy
Successfully installed iniconfig-2.0.0 pluggy-1.5.0 pytest-8.2.2 swcpy-0.0.1
```

The `pip` package adds the version number on your Python package from the *py project.toml* file.

Congratulations, you have created and installed a Python package! Take a look at the files you have created for this minimal SDK. Use the `tree` command, with a few options to filter out temporary files created by the build process:

```
.../sdk (main) $ tree --prune -I 'build|*.egg-info|__pycache__'
.
├── pyproject.toml
├── src
│   └── swcpy
│       ├── __init__.py
│       └── swc_client.py


2 directories, 3 files
```

Minimum Viable SDK as a Building-Block Skill

The code you have created at this point in the chapter is deceptively simple, only requiring two directories and three files. But as basic as this appears, consider that you have now learned the basics required to:

- Create a simple SDK wrapper for any API that is available on the internet.
- Make a package from any Python library and install it using `pip3`. This will work for any Python code that you think is useful to share.

As you will see later in the chapter, this package structure will allow you to share these files with other users on GitHub and install it in their environments using `pip3`. With a bit more effort, you can also publish this package on PyPI.

You have picked up a key building-block skill that you can use as a foundation to share your work with the community of data scientists and developers around the world.

Building a Feature-Rich SDK

Now it's time to turn the minimum viable SDK into one that is robust and feature rich. As you learn expert tips, you can implement them in your code. The goal is to make your SDK *batteries included*, which means it comes with all the functionality that users need for interacting with your API. This is where an SDK becomes a major selling point to your users, and it gives them the ability to get the most from your API.

Expert Tip: Using Sane Defaults

A key to hiding complicated details is to implement sane defaults. This means that a user can use your SDK without specifying any parameters and it will work out of the box. One important default is that the SDK should know the base URL of the API without being told. This allows the user to `pip3` install it and use it without reading the documentation to know this address. If the SDK is a read-only wrapper for a public API, you'll likely default the production API address. If the API requires authentication or is a read/write API, you may want to default to a sandbox environment to prevent accidents. The sane defaults will handle the *happy path*, the standard usage that 80% of your users will need. The remaining 20% can be handled by allowing the user to change the configuration to override the defaults for special situations they have.

You will add sane defaults that the user can override by creating a configuration class. Create a file named *swc_config.py* as shown:

```
.../sdk (main) $ touch src/swcpy/swc_config.py
```

This file defines the `SWCConfig` class. The user will instantiate an instance of this class with their configuration settings, and then pass it to the `SWCClient` constructor. Then, they will use the `SWCClient` to call the API.

Update *swc_config.py* with the following content:

```
import os
from dotenv import load_dotenv  ❶

load_dotenv()  ❷

class SWCConfig:
    """Configuration class containing arguments for the SDK client.

    Contains configuration for the base URL and progressive backoff.
    """

    swc_base_url: str  ❸
    swc_backoff: bool  ❹
    swc_backoff_max_time: int  ❺
    swc_bulk_file_format: str  ❻

    def __init__(  ❼
        self,
        swc_base_url: str = None,
        backoff: bool = True,
        backoff_max_time: int = 30,
        bulk_file_format: str = "csv",
    ):
        """Constructor for configuration class.
```

```
    Contains initialization values to overwrite defaults.

    Args:
    swc_base_url (optional):
        The base URL to use for all the API calls. Pass this in or set
        in environment variable.
    swc_backoff:
        A boolean that determines if the SDK should
        retry the call using backoff when errors occur.
    swc_backoff_max_time:
        The max number of seconds the SDK should keep
        trying an API call before stopping.
    swc_bulk_file_format:
        If bulk files should be in csv or parquet format.
    """

    self.swc_base_url = swc_base_url or os.getenv("SWC_API_BASE_URL")  ❽
    print(f"SWC_API_BASE_URL in SWCConfig init: {self.swc_base_url}")


    if not self.swc_base_url:  ❾
        raise ValueError("Base URL is required. Set SWC_API_BASE_URL
                          environment variable.")

    self.swc_backoff = backoff
    self.swc_backoff_max_time = backoff_max_time
    self.swc_bulk_file_format = bulk_file_format

def __str__(self):  ❿
    """Stringify function to return contents of config object for logging"""
    return f"{self.swc_base_url} {self.swc_backoff}
    {self.swc_backoff_max_time}  {self.swc_bulk_file_format}"
```

❶ This import will be used to get environment variables from the Python environment.

❷ This statement loads external variables from a *.env* file or the operating systems environment.

❸ The swc_base_url will be used to access the API.

❹ The swc_backoff determines if the SDK should retry the call using backoff when errors occur.

❺ The swc_backoff_max_time The max number of seconds the SDK should keep trying an API call before stopping.

❻ The swc_bulk_file_format sets the format for bulk files.

❼ The `__init__` method is executed onced when this class is created. It is used to set the class variables from the parameters the user passes in. Default values are set in this method.

❽ This line sets the internal URL from a parameter that passed in the constructor, or from the environment if there is no parameter in the constructor.

❾ This statement checks if a URL has been provided through one of the methods described previously. If none is present, this raises an exception and the loading stops. This class can't be used without a URL to access the API.

❿ The `__str__` method is used by external programs to log the contents of this class. If you don't provide a custom method like this, a default method would be created, but it might have less useful information.

Next, modify the *__init__.py* file in the *chapter7/sdk/src/swcpy* directory so that it looks like the following:

```
from .swc_client import SWCClient
from .swc_config import SWCConfig
```

These imports simplify the process of importing the classes in a user's code.

Expert Tip: Providing Rich Functionality

Simon Yu suggested that an SDK should be rich with features that save a significant amount of coding for developers. A survey of the SDKs I reviewed in this chapter found features such as handling versions, handling pagination, client-side caching, and authentication. You will include two advanced features: data type validation and retry/backoff logic.

It is time-consuming for end users to add data validation in their code. They have to read the API documentation to determine valid data types and values, and then add a lot of code that checks values and throws errors. When SDK developers add data validation, this is a big benefit to users. The original API developers have detailed knowledge of the workings of the API, which makes it easier for them to add validations to the SDK. You have an additional advantage for this Python SDK: access to the original Pydantic classes the API was built with. You can reuse the Chapter 7 *schemas.py* file in your SDK, giving your SDK powerful data validation at a minimal effort.

Create a *schemas* folder and *__init__.py* file as follows:

```
.../sdk (main) $ mkdir src/swcpy/schemas
.../sdk (main) $ echo "from .schemas import *" > src/swcpy/schemas/__init__.py
```

Copy the schemas file from the *chapter6* directory (or *chapter6/complete* if you haven't finished Chapter 6) into this chapter's directories as follows:

```
.../sdk (main) $ cp ../../chapter6/complete/schemas.py src/swcpy/schemas
```

Now you will rebuild the client from scratch, step by step. First, you will add the imports and the class constructor. Replace the entire *swc_client.py* file with the following content:

```python
import httpx
import swcpy.swc_config as config  ❶
from .schemas import League, Team, Player, Performance  ❷
from typing import List  ❸
import backoff  ❹
import logging  ❺
logger = logging.getLogger(__name__)

class SWCClient:
    """Interacts with the SportsWorldCentral API.

        This SDK class simplifies the process of using the SWC Fantasy
        Football API. It supports all the functions of the SWC API and returns
        validated data types.

    Typical usage example:

        client = SWCClient()
        response = client.get_health_check()

    """

    HEALTH_CHECK_ENDPOINT = "/"  ❻
    LIST_LEAGUES_ENDPOINT = "/v0/leagues/"
    LIST_PLAYERS_ENDPOINT = "/v0/players/"
    LIST_PERFORMANCES_ENDPOINT = "/v0/performances/"
    LIST_TEAMS_ENDPOINT = "/v0/teams/"
    GET_COUNTS_ENDPOINT = "/v0/counts/"

    BULK_FILE_BASE_URL = (
        "https://raw.githubusercontent.com/[github ID]"  ❼
        + "/portfolio-project/main/bulk/"
    )

    def __init__(self, input_config: config.SWCConfig):  ❽
        """Class constructor that sets variables from configuration object."""

        logger.debug(f"Bulk file base URL: {self.BULK_FILE_BASE_URL}")

        logger.debug(f"Input config: {input_config}")

        self.swc_base_url = input_config.swc_base_url
        self.backoff = input_config.swc_backoff
```

```
        self.backoff_max_time = input_config.swc_backoff_max_time
        self.bulk_file_format = input_config.swc_bulk_file_format

        self.BULK_FILE_NAMES = { ❾
            "players": "player_data",
            "leagues": "league_data",
            "performances": "performance_data",
            "teams": "team_data",
            "team_players": "team_player_data",
        }

        if self.backoff: ❿
            self.call_api = backoff.on_exception(
                wait_gen=backoff.expo,
                exception=(httpx.RequestError, httpx.HTTPStatusError),
                max_time=self.backoff_max_time,
                jitter=backoff.random_jitter,
            )(self.call_api)

        if self.bulk_file_format.lower() == "parquet": ⓫
            self.BULK_FILE_NAMES = {
                key: value + ".parquet" for key, value in
                self.BULK_FILE_NAMES.items()
            }
        else:
            self.BULK_FILE_NAMES = {
                key: value + ".csv" for key, value in
                self.BULK_FILE_NAMES.items()
            }

        logger.debug(f"Bulk file dictionary: {self.BULK_FILE_NAMES}")
```

❶ This imports the *swc_config.py* file you created.

❷ This is used to import the Pydantic schemas.

❸ This is used for additional type hints for the Pydantic classes.

❹ This is used to implement exponential backoff.

❺ This import statement and the following line of code are used to log debug and error messages.

❻ This adds class constants for all the API's endpoints.

❼ You need to replace the *[github ID]* with your GitHub ID so that the path to the bulk files works correctly.

❽ The SWCClient class constructor accepts an instance of SWCConfig now. The user puts all of their configuration in this object and passes it to the client.

❾ This creates a dictionary of bulk filenames without a file extension.

❿ This is a conditional decorator, which will update the call_api() method to have backoff functionality if the user configures it. The backoff will be explained further later.

⓫ This code appends a file extension of *.csv* or *.parquet* to the filenames dictionary. It uses a dictionary comprehension, which is an efficient and Pythonic way to update all the elements in a dictionary.

The retry and backoff functionality you added in the client has a few unexpected twists and turns. When you make an API call from your SDK, sometimes it may fail due to a temporary network hiccup or slowdown in the API. It might be due to a load-balancing issue or a service that is in the middle of bringing more servers online to handle increased load. If you are running a process that uses the API, you can make it more resilient by retrying a few times before giving up.

However, the way you retry may cause some unexpected consequences on the API provider. For example, you could decide that if an API call fails, you will continually retry every microsecond until it succeeds. This would be the equivalent of a web user clicking Refresh continually until the web page is back up. One user doing that might not cause that much of a headache. But what about when hundreds or thousands of users are retrying the API continually in the middle of a network outage, or when an API provider is trying to bring more resources online? These users will be hitting it with a massive load at a time when it's least able to respond. This is like an accidental distributed denial-of-service (DDOS) attack from your own users.

API consumers and SDK developers can implement *exponential backoff* instead of simple retries. With this method, the time between retries gets exponentially longer with each failed attempt. This method recognizes that a service that is failing needs a break, and the more failures that happen, the more of a break it needs. For example, a client might double the time between retries each time it fails. Using this method, it would retry at 1 second, 2 seconds, 4 seconds, 8 seconds, and so on until it reached a preset stopping point.

This is an improvement, but can you see any issue with everyone using this method? The problem is explained very clearly on AWS's *Architecture* blog in "Exponential Backoff And Jitter" (*https://oreil.ly/tSRCm*), and Figure 7-3 gives a simplified visualization. The numbers across the bottom are seconds after an outage, and the vertical bars are the number of retry attempts occurring in that second.

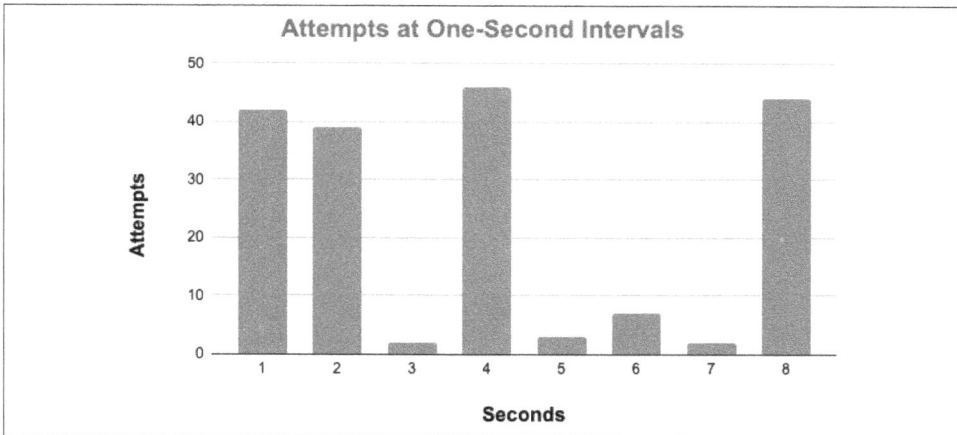

Figure 7-3. The problem with exponential backoff

As you can see from the image, if multiple people are using the same settings for exponential backoff (as they would be if they're all using your SDK with defaults), they are likely to still cluster around the same intervals to retry. This is because, when an outage occurs, it will probably hit many users at the same time, starting their retry clock. This is not what you want. The solution is to introduce *jitter*, which is a some‐ what random element that is combined with the backoff. Using this method, you get the benefit of backing off retries without the clustering of retries in the same inter‐ vals. The best of both worlds is exponential backoff with jitter.

The code you added to *swc_client.py* implemented exponential backoff with random jitter using the backoff Python library (*https://oreil.ly/PUZbE*). To use this library, you added the decorator `backoff.on_exception` onto the `call_api()` function. This wraps the function call with backoff, without having to make any changes to the func‐ tion itself. If the SDK user uses backoff, the `call_api()` function will be wrapped with this additional functionality.

Take another look at the decorator you added to your code:

```
if self.backoff:
    self.get_url = backoff.on_exception(
        wait_gen=backoff.expo, ❶
        exception=(httpx.RequestError, httpx.HTTPStatusError), ❷
        max_time=self.backoff_max_time, ❸
        jitter=backoff.random_jitter, ❹
    )(self.call_api)
```

❶   If an exception occurs, retry starting at 1 second, then doubling every retry.

❷   Look for `RequestError` and `HTTPStatusError` exceptions from the `call_api` function.

❸ Stop retrying after the `backoff_max_time` parameter (which defaults to 30 seconds).

❹ Apply random jitter so that it varies slightly from the exact second.

Expert Tip: Performing Logging

You don't want your SDK to be a black box—you want users to understand what's occurring under the hood. One way to do this is by publishing the source code on a public repository. The other is by providing meaningful logging, which helps when the user is encountering errors or isn't returning the results expected.

To be Pythonic, your SDK will use Python's built-in *logging* library, and allow the users to configure what level of logging they want to print. For example, they may only want to log serious errors in production logs while logging debug messages during development. Table 7-1 shows the logging levels available, according to the official Python docs (*https://oreil.ly/CmUps*).

Table 7-1. Python logging levels

Level	What it means/when to use it
`logging.DEBUG`	Detailed information, typically only of interest to a developer trying to diagnose a problem.
`logging.INFO`	Confirmation that things are working as expected.
`logging.WARNING`	An indication that something unexpected happened, or that a problem might occur in the near future (e.g., disk space low). The software is still working as expected.
`logging.ERROR`	Indicates that, due to a more serious problem, the software has not been able to perform some function.
`logging.CRITICAL`	A serious error, indicating that the program itself may be unable to continue running.

> In this chapter, you will see logging messages in the terminal when you run `pytest` if an error occurs. By default, any messages with a log level of `WARNING` or more severe will be displayed. If you want to see `INFO` or `DEBUG` messages, you can use the `--log-level` command-line option.

You added the import statements and created a `logger` object earlier in this chapter. You will see this functionality in action in a new method that you will create, named `call_api()`. Add the following code at the bottom of *swc_client.py*:

```
def call_api(self, ❶
        api_endpoint: str, ❷
        api_params: dict = None ❸
    ) -> httpx.Response: ❹
    """Makes API call and logs errors."""
```

```
        if api_params: ❺
            api_params = {key: val for key, val in api_params.items() if val
            is not None}

        try:
            with httpx.Client(base_url=self.swc_base_url) as client:  ❻
                logger.debug(f"base_url: {self.swc_base_url}, api_endpoint:
                {api_endpoint}, api_params: {api_params}") ❼
                response = client.get(api_endpoint, params=api_params)
                logger.debug(f"Response JSON: {response.json()}")
                return response
        except httpx.HTTPStatusError as e:
            logger.error(
                f"HTTP status error occurred: {e.response.status_code}
                {e.response.text}"
            )
            raise
        except httpx.RequestError as e:
            logger.error(f"Request error occurred: {str(e)}")
            raise
```

❶ Because this is a class method, the first parameter is always `self`, which represents the instance of this class.

❷ The second parameter to this method is the individual endpoint you are calling.

❸ The query string parameters for the API call are passed in as an optional dictionary.

❹ This is a type hint that the method should return an `httpx.Response` object.

❺ This dictionary comprehension removes any empty parameters before calling the API with them.

❻ This is a context manager that uses `httpx.Client` for the steps that follow.

❼ This is logging at the `logging.DEBUG` level. The parameters are formatted with an F-string, which is a Pythonic way of formatting variable values.

The `call_api` function is used to make the API calls for each SDK function. By centralizing this work, you can apply additional error handling and logging without making your code too long. It adds `try...except` logic around the API call. If the API call works, this function logs a `logging.DEBUG` level message with the data from the response.

Let's take a closer look at a key line of code in this method:

```
with httpx.Client(base_url=self.swc_base_url) as client:
```

When a statement uses the with…as format like this, it is using a Python object as a *context manager*, so that the object is initialized, then the statements inside it run, and then it is cleaned up along with any resources it used. Here, the httpx.Client is the context manager.

According to the HTTPX documentation (*https://oreil.ly/I-xln*), the httpx.Client is an efficient way to make API calls using *httpx* because it pools resources between API calls. The Client constructor accepts the parameter base_url and will use it for all of the API calls it is used for.

This code is contained inside at try…except block, so if the API call doesn't work, the except first handles httpx.HTTPStatusError, which is a type of error that will have an HTTP status code. For this type of exception, it logs a logging.ERROR level message and then re-raises the exception. If it's not that type of exception, it handles httpx.RequestError next. This type of exception doesn't have an HTTP status code, so it just puts the body of the exception in the log message. Then it re-raises the exception. Re-raising the exception is important because you will be adding retry and backoff logic later in this chapter, and it will be looking for those exceptions.

Expert Tip: Hiding Your API's Complicated Details

Joey Greco likes SDKs to handle some of the complexities of the underlying API. "As a data consumer, I don't want to have to worry about API versioning, headers, authentication, rate-limiting, or hunting down the correct endpoints to use," he said. "I just want to call some function and be able to do things that are meaningful to me."

Now add the get_health_check and list_leagues methods to the bottom of the file. Endpoint methods like get_health_check and list_leagues wrap the API calls and shield users from complicated details such as endpoint names and return types. It's not an exaggeration to say that methods like these are the reason that SDKs exist.

These both use the call_api function that you created. The list_leagues method also uses Pydantic to validate the data returned from the API. If the data from the API does not match the Pydantic class definition, the client will throw an error that can be logged. Add the following code to the bottom of the *swc_client.py* file:

```
def get_health_check(self) -> httpx.Response:
    """Checks if API is running and healthy.

    Calls the API health check endpoint and returns a standard
    message if the API is running normally. Can be used to check
    status of API before making more complicated API calls.

    Returns:
      An httpx.Response object that contains the HTTP status,
      JSON response and other information received from the API.
```

```
        """
        logger.debug("Entered health check")
        endpoint_url = self.HEALTH_CHECK_ENDPOINT ❶
        return self.call_api(endpoint_url)

    def list_leagues(
        self,
        skip: int = 0,
        limit: int = 100,
        minimum_last_changed_date: str = None,
        league_name: str = None,
    ) -> List[League]: ❷
        """Returns a List of Leagues filtered by parameters.

        Calls the API v0/leagues endpoint and returns a list of
        League objects.

        Returns:
        A List of schemas.League objects. Each represents one
        SportsWorldCentral fantasy league.

        """
        logger.debug("Entered list leagues")

        params = {  ❸
            "skip": skip,
            "limit": limit,
            "minimum_last_changed_date": minimum_last_changed_date,
            "league_name": league_name,
        }

        response = self.call_api(self.LIST_LEAGUES_ENDPOINT, params) ❹
        return [League(**league) for league in response.json()] ❺
```

❶ The new get_health_check() method uses the call_api() method, instead of calling the API directly as it did in the minimal SDK.

❷ The type hint shows that the method should return a List of League objects. League is a Pydantic class defined in the *schemas.py* file.

❸ The parameters will be passed in when users call this SDK method. This line of code builds a dictionary containing the parameters.

❹ This method also calls the call_api method and passes in the query string parameters as a dictionary.

❺ This looks through the list of dictionaries returned in the API response and populates a List of League objects.

Take a look at the final line of code in list_leagues(), which packs a lot into a short syntax:

```
return [League(**league) for league in response.json()]
```

Your goal in this statement is to iterate through the list of dictionaries returned from the API and create a list of Pydantic League objects. You use a list comprehension (*https://oreil.ly/Fpgem*), which is a Pythonic way to build lists without using a recursive loop. Using the general syntax list = [expression for item in iterable], you can create lists from other lists very easily.

As you iterate through, you want to create a Pydantic League object from each dictionary in the original list. You can do this with Python's *unpacking operator*. The statement League(**league) uses two asterisks to unpack the original dictionary into key-value pairs that are passed to the League() constructor. Pydantic performs data validation during the process of creating these objects, so if any invalid data is in the API response, it will error out in this step. With this combination of list comprehension and the unpacking operator, you return a list of League objects that have been validated from the original list of dictionaries.

Expert Tip: Supporting Bulk Downloads

> As a data scientist, I find the complexity of many open data services frustrating. I don't want to have to learn how to query an endpoint, think about data types, or read through API documentation. Just give me the data!
>
> —Robin Linacre

Bulk downloads are valuable for many API users, but data scientists especially like this feature. Data scientists often use full datasets for analysis and loading data, and they find it frustrating to call multiple endpoints to get subsets of the data. You could serve bulk downloads from an endpoint in your API using FastAPI's Static Files (*https://oreil.ly/58q0a*).

Instead, you will be serving bulk downloads from SDK methods. Your SDK will access the files from their web-hosted location in your GitHub repository. The URL for each file will be built using the BULK_FILE_BASE_URL and the BULK_FILE_NAMES dictionaries. Your repository contains two bulk files for each table you loaded in your SQLite database: one in *.csv* format and one in *.parquet* format. By providing these two options, your SDK serves a wide range of bulk data needs.

The bulk files are in the *bulk* folder of your repository. View the list of these files with the following commands:

```
.../sdk (main) $ ls /workspaces/portfolio-project/bulk
csv_to_parquet.py        performance_data.parquet  team_data.csv
league_data.csv          player_data.csv           team_data.parquet
league_data.parquet      player_data.parquet       team_player_data.csv
performance_data.csv     readme.md                 team_player_data.parquet
```

The files with the *.csv* extension are *comma-separated values* files. The following shows the first two rows of the *player_data.csv* file:

```
player_id,gsis_id,first_name,last_name,position,last_changed_date
1001,00-0023459,Aaron,Rodgers,QB,2024-04-18
```

These are plain-text files with no compression of any kind. The first row contains the column names separated by commas. The remaining rows contain one row for each data record and the data values are separated by commas. CSV files are easily processed in Python libraries such as pandas or in software such as Microsoft Excel.

Parquet files use an open source data format that is popular for a variety of data analysis tasks. Here is the official definition from the Apache Parquet project page (*https://oreil.ly/K4_0_*): "Apache Parquet is an open source, column-oriented data file format designed for efficient data storage and retrieval. It provides high performance compression and encoding schemes to handle complex data in bulk and is supported in many programming language and analytics tools."

You will create a separate method to retrieve each file, but the method will retrieve either *.csv* or *.parquet* format files based upon the options provided in the bulk_file_format parameter the user selects in the SWCConfig class. The default value is *.csv* if the user does not provide this parameter.

Add the following method at the bottom of *swc_client.py*:

```
def get_bulk_player_file(self) -> bytes: ❶
    """Returns a bulk file with player data"""

    logger.debug("Entered get bulk player file")

    player_file_path = self.BULK_FILE_BASE_URL + self.BULK_FILE_NAMES
                    ["players"] ❷

    response = httpx.get(player_file_path, follow_redirects=True) ❸

    if response.status_code == 200:
        logger.debug("File downloaded successfully")
        return response.content ❹
```

❶ The type hint for this method is `bytes` because the Parquet file is a binary file, not text.

❷ This statement builds the URL to an individual file from the web-hosted location in your GitHub repository using the `BULK_FILE_BASE_URL` and `BULK_FILE_NAMES` dictionaries.

❸ Instead of using the `call_api` method, like the other methods do, this method uses the `httpx.get()` method. The `follow_redirects` parameter handles any web redirects that occur when retrieving the file.

❹ The method returns `response.content`, which will contain the binary file.

You have created methods that implement all the major functionality of your SDK, although you haven't created methods for all the endpoints or bulk files yet. Update your *pyproject.toml* file to contain all the new libraries you've added, and increment your version number:

```
[build-system]
requires = ["setuptools>=61.0"]
build-backend = "setuptools.build_meta"

[project]
name = "swcpy"
version = "0.0.2" ❶
authors = [
  { name="[enter your name]" },
]
description = "Example Software Development Kit (SDK) from Hands-On API Book
for AI and Data Science"
requires-python = ">=3.10"
classifiers = [
    "Development Status :: 3 - Alpha",
    "Programming Language :: Python :: 3",
    "License :: OSI Approved :: MIT License",
    "Operating System :: OS Independent",
]
dependencies = [
        'pytest>=8.1',
        'httpx>=0.27.0',
        'pydantic>=2.4.0', ❷
        'backoff>=2.2.1', ❸
        'pyarrow>=16.0', ❹
]
```

❶ Increment the version number of the library to reflect all the new functionality.

❷ Import the Pydantic library for your data validation.

---

❸ Import the backoff library for your backoff and retry functionality.

❹ Import the PyArrow library for handling Parquet files.

Expert Tip: Documenting Your SDK

Just as APIs need documentation, so does your SDK. Since your SDK will be used directly in program code, the first step to documenting your SDK is to add comprehensive docstrings to the methods that will be used by programmers. This is an important part of writing Pythonic code, and it helps data scientists using your SDK in an IDE like VS Code to get hints and code completion that make their work faster and more accurate. As more data scientists use generative AI in their development process, it allows AI assistants and copilots to infer accurate coding suggestions. You provided extensive docstrings in the *swc_client.py* and *swc_config.py* files.

You also need to include a well-written *README.md* file that explains how to install the SDK and provides examples of using it. This file will be displayed by default in the GitHub repository for your SDK, and it will be the home page for your SDK if you publish it to PyPI.

Create the *README.md* file as follows:

```
.../sdk (main) $ touch README.md
```

Add the following contents to this file:

```
# swcpy software development kit (SDK)
This is the Python SDK to interact with the SportsWorldCentral Football API,
which was created for the book [Hands-On APIs for AI and Data Science]
(https://handsonapibook.com).

## Installing swcpy

To install this SDK in your environment, execute the following command:

`pip install swcpy@git+https://github.com/{owner of repo}/
 portfolio-project#subdirectory=sdk`

## Example usage

This SDK implements all the endpoints in the SWC API, in addition to providing
bulk downloads of the SWC fantasy data in CSV format.

### Setting base URL for the API
The SDK looks for a value of `SWC_API_BASE_URL` in the environment. The preferred
method for setting the base URL for the SWC API is by creating a Python
`.env` file in your project directory with the following value:

```
SWC_API_BASE_URL={URL of your API}
```
```

```
```

You may also set this value as an environment variable in the environment you are using the SDK, or pass it as a parameter to the `SWCConfig()` method.

Example of normal API functions

To call the SDK functions for normal API endpoints, here is an example:

```python
from swcpy import SWCClient
from swcpy import SWCConfig

config = SWCConfig(swc_base_url="http://0.0.0.0:8000",backoff=False)
client = SWCClient(config)
leagues_response = client.list_leagues()
print(leagues_response)
```

Example of bulk data functions

The build data endpoint returns a bytes object. Here is an example of saving a file locally from a bulk file endpoint:

```python
import csv
import os
from io import StringIO

config = SWCConfig()
    client = SWCClient(config)

    """Tests bulk player download through SDK"""
    player_file = client.get_bulk_player_file()


    # Write the file to disk to verify file download
    output_file_path = data_dir + 'players_file.csv'
    with open(output_file_path, 'wb') as f:
        f.write(player_file)
```

One key section to notice in this file is that it explains to users how they can install your SDK from your GitHub repository. Here is that section:

Installing swcpy

To install this SDK in your environment, execute the following command:

`pip install swcpy@git+https://github.com/{owner of repo}/portfolio-project#subdirectory=sdk`

You have now created all the files you need and completed all of the coding for the first few endpoints. To see the new structure of your project, run the `tree` command:

```
.../sdk (main) $ tree --prune -I 'build|*.egg-info|__pycache__'
.
├── README.md
├── pyproject.toml
└── src
    └── swcpy
        ├── __init__.py
        ├── schemas
        │   ├── __init__.py
        │   └── schemas.py
        ├── swc_client.py
        └── swc_config.py

3 directories, 7 files
```

To update your local system with the new version of the SDK using `pip`, you will use the `--upgrade` option:

```
.../sdk (main) $ pip3 install --upgrade .
Processing /workspaces/portfolio-project/chapter7/sdk
  Installing build dependencies ... done
  Getting requirements to build wheel ... done
  Preparing metadata (pyproject.toml) ... done
...
Successfully built swcpy
Installing collected packages: pydantic-core, backoff, annotated-types, pydantic,
swcpy
  Attempting uninstall: swcpy
    Found existing installation: swcpy 0.0.1
    Uninstalling swcpy-0.0.1:
      Successfully uninstalled swcpy-0.0.1
Successfully installed annotated-types-0.7.0 backoff-2.2.1 pyarrow-16.1.0
pydantic-2.4.2 pydantic-core-2.10.1 swcpy-0.0.2
```

Testing Your SDK

Now you will test the SDK with pytest. Create a new directory and file with the following commands:

```
.../sdk (main) $ mkdir tests
.../sdk (main) $ touch tests/test_swcpy.py
```

> There are different pytest layouts you can use to include the tests in your package. In this case, you are using the pytest layout style named "tests outside application" that is described in pytest's Good Integration Practices (*https://oreil.ly/GM3HU*). This means that when you run your tests, you are testing against the installed module instead of code in your local path.

Update *test_swcpy.py* with the following contents:

```python
import pytest
from swcpy import SWCClient ❶
from swcpy import SWCConfig
from swcpy.schemas import League, Team, Player, Performance
from io import BytesIO ❷
import pyarrow.parquet as pq ❸
import pandas as pd ❹

def test_health_check(): ❺
    """Tests health check from SDK"""
    config = SWCConfig(swc_base_url="http://0.0.0.0:8000",backoff=False)
    client = SWCClient(config)
    response = client.get_health_check()
    assert response.status_code == 200
    assert response.json() == {"message": "API health check successful"}

def test_list_leagues(): ❻
    """Tests get leagues from SDK"""
    config = SWCConfig(swc_base_url="http://0.0.0.0:8000",backoff=False)
    client = SWCClient(config)
    leagues_response = client.list_leagues()
    # Assert the endpoint returned a list object
    assert isinstance(leagues_response, list)
    # Assert each item in the list is an instance of a Pydantic League object
    for league in leagues_response:
        assert isinstance(league, League)
    # Asset that 5 League objects are returned
    assert len(leagues_response) == 5

def test_bulk_player_file_parquet(): ❼
    """Tests bulk player download through SDK - Parquet"""

    config = SWCConfig(bulk_file_format = "parquet") ❽
    client = SWCClient(config)

    player_file_parquet = client.get_bulk_player_file()

    # Assert the file has the correct number of records (including header)
    player_table = pq.read_table(BytesIO(player_file_parquet)) ❾
    player_df = player_table.to_pandas()
    assert len(player_df) == 1018
```

❶ The import statement is referencing the package that you installed locally in your environment.

❷ This library is used for handling binary files like the Parquet file.

❸ This library is specifically used to process Parquet files.

❹ You will use the pandas library to read the length of the Parquet files.

❺ This test method tests health check endpoints.

❻ This test method tests the method calling your API.

❼ This test method tests the Parquet bulk file download.

❽ This sets the configuration option for Parquet files.

❾ These lines of code use PyArrow and pandas to read the Parquet file and count the records.

Now you will run your API in another terminal session so that your SDK can call it. Open a second terminal session in Codespaces using the split terminal command, as shown in Figure 7-4.

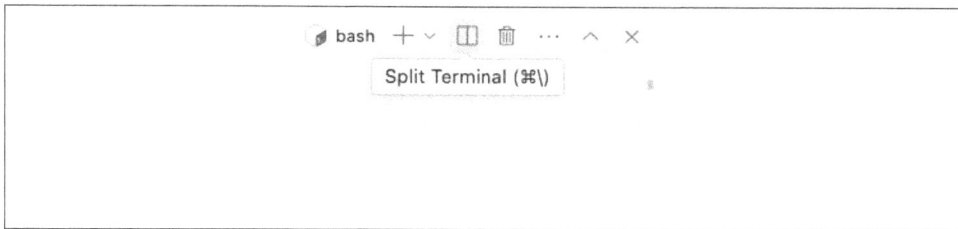

Figure 7-4. Opening a second terminal session

When you have added the second terminal, you should see it in the split screen, as shown in Figure 7-5.

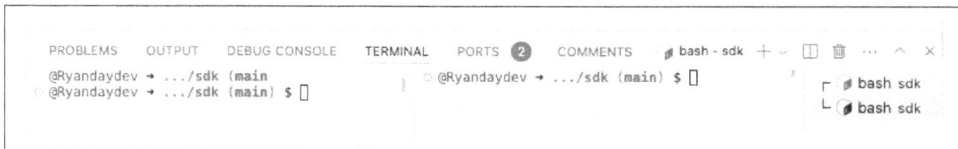

Figure 7-5. The split terminal session

If you haven't installed the API in your Codespace previously, you will need to run the command `pip3 install -r requirements.txt` in the *chapter6/complete* directory before running the API.

In the second window, change directories to *chapter6/complete* to use the completed API from the repository. Launch the API as shown:

```
.../sdk (main) $ cd ../../chapter6/complete
.../chapter6 (main) $ fastapi run main.py
...
INFO:      Started server process [9999]
INFO:      Waiting for application startup.
INFO:      Application startup complete.
INFO:      Uvicorn running on http://0.0.0.0:8000 (Press CTRL+C to quit) ❶
```

❶ The *test_swcpy.py* file should use this address in the swc_base_url= parameter. If it doesn't have that address, update *test_swcpy.py* to match the address here.

You don't need to click "Open in Browser" as in previous chapters, because you will be testing the API from the terminal using your SDK.

In the left terminal window, enter the **pytest tests/test_swcpy.py** command and you should see an output that looks similar to this:

```
.../sdk (main) $ pytest tests/test_swcpy.py
=========== test session starts ===========
platform linux -- Python 3.10.13, pytest-8.1.2, pluggy-1.5.0
rootdir: /workspaces/portfolio-project/chapter7/sdk
configfile: pyproject.toml
plugins: anyio-4.4.0
collected 3 items

tests/test_swcpy.py ...                    [100%]

============ 3 passed in 0.68s ============
```

You have added a lot of great functionality to your SDK. Way to go! Take a moment to consider expert tips and features that you have implemented for your SDK:

Make your SDK easy to install.
    You configured your project so that it can be installed using pip3 directly from GitHub. You could also publish it to PyPI if you choose.

Be consistent and idiomatic.
    You used PEP 8 style and consistent function naming. You used Pythonic techniques like list comprehensions, dictionary comprehensions, and context managers. And you used the Python standard logging function.

Use sane defaults.
    You implemented the SWCConfig that will work out of the box with default values but can be customized with URL and other settings.

Provide extra functionality.
> You provided data validation, exponential backoff with jitter, and bulk downloads.

Perform logging.
> You implemented logging with multiple levels using Python's built-in logging library.

Hide your API's complicated details.
> You implemented methods that allow users to call the API endpoints without reading the API documentation.

Provide bulk downloads.
> You made bulk files available for all your database tables in both *.csv* and *.parquet* formats.

Document your SDK.
> You provided a *README.md* file that explains how to install and use the SDK.

Only one expert tip remains for your SDK, but it's a big one. You will tackle it next.

Expert Tip: Supporting Every Task the API Supports

Ideally, users should be able to accomplish any task with the SDK that they could accomplish by directly using the API. This means every API endpoint and parameter should be supported by the SDK. You've probably noticed at this point that you've only implemented SDK functions for a couple of the API endpoints and one bulk file download. Unfortunately, there's no room in this chapter to walk you through the remaining code. However, you are prepared to code the rest of the endpoints using the helper functions and naming standards I have demonstrated so far.

I don't want to leave you hanging completely, so here is some additional information you can use to continue building your SDK. To start with, here are the functions you need to create to cover all the API endpoints:

- get_health_check (completed)
- list_leagues (completed)
- get_league_by_id
- get_counts
- list_teams
- list_players
- get_player_by_id
- list_performances

Here are the bulk download functions you need to create:

- get_bulk_player_file (completed)
- get_bulk_league_file
- get_bulk_performance_file
- get_bulk_team_file
- get_bulk_team_player_file

I encourage you to take a shot at creating these endpoints following the examples I have given you. Be consistent and idiomatic. As with earlier chapters, the full completed code is available in the *chapter7/complete* directory if you would like to check your work.

Completing Your Part I Portfolio Project

With the creation of your SDK, you have finished all the coding you need for your portfolio project. Congratulations! Hopefully, you have been committing your code with frequent small commits, but be sure to commit any remaining changes you have.

To get your project repository ready to share, you are going to be moving the Chapters 6 and 7 content to the root folder of your repository, and then removing all the previous chapters.

Before you make these changes, you'll save a copy of your files to a separate GitHub *branch*, which is an isolated area in your repository. This will keep the original directory structure you used while working through your code. (You have been doing all your coding in the *main* branch so far.) Create the new branch from the command line as follows:

```
.../sdk (main) $ cd ../..  ❶
.../portfolio-project/ (main) $ git checkout -b learning-branch  ❷
Switched to a new branch 'learning-branch'
.../portfolio-project/ (main) $ git push -u origin learning-branch  ❸
 * [new branch]      learning-branch -> learning-branch
branch 'learning-branch' set up to track 'origin/learning-branch'.
```

❶ Change to the root directory.

❷ Create a new branch named *learning-branch* locally based on the *main* branch.

❸ Push this new branch to your remote repository on GitHub.com.

To verify that the branch was created on GitHub, go to your GitHub repository and click main. You should see a new branch, as shown in Figure 7-6.

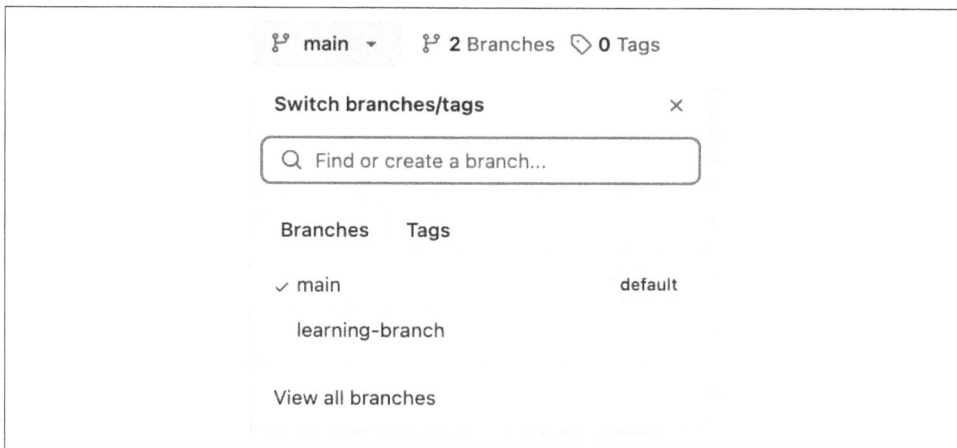

Figure 7-6. Newly created branch

Back in Codespaces, you will move the Chapters 6 and 7 files to the root of your repository. Chapter 6 contains the final API files and Chapter 7 has the SDK files. Enter the following commands:

```
.../portfolio-project/ (learning-branch) $ git checkout main ❶
Switched to branch 'main'
Your branch is up to date with 'origin/main'.
.../portfolio-project/ (main) $ rm -rf chapter6/complete
.../portfolio-project/ (main) $ rm -rf chapter7/complete
.../portfolio-project/ (main) $ rm
.../portfolio-project/ (main) $ mv chapter6/* . ❷
.../portfolio-project/ (main) $ mv chapter7/sdk . ❸
.../portfolio-project/ (main) $ rm -rf chapter3 ❹
.../portfolio-project/ (main) $ rm -rf chapter4
.../portfolio-project/ (main) $ rm -rf chapter5
.../portfolio-project/ (main) $ rm -rf chapter6
.../portfolio-project/ (main) $ rm -rf chapter7
```

❶ Switch your Codespace back to the *main* branch of your repository.

❷ Move the API files into your root directory.

❸ Move the SDK files into your *sdk* directory.

❹ Remove all the subdirectories and their files.

To see the directory structure of the completed project, run the following command:

```
.../portfolio-project (main) $ tree -d  --prune -I 'build|*.egg-info|__pycache__'
.
├── bulk
└── sdk
    ├── src
    │   └── swcpy
    │       └── schemas
    └── tests
6 directories
```

To update your API documentation to mention your SDK, replace the bottom "Software Development Kit (SDK)" section of *README.md* with the following:

```
## Software Development Kit (SDK)

If you are a Python user, you can use the swcpy SDK to interact with our API.
Full information is available [here](sdk/README.md).
```

You've performed some serious surgery to your *main* branch. Make one last commit to GitHub, and you're all done. Congratulations on completing your Part I capstone!

Additional Resources

For advice on creating an SDK, see the SDKs.io website (*https://sdks.io*) by APIMatic.

For more advice from Speakeasy on creating a Python SDK, read "How to Build a Best in Class Python SDK" by Tristan Cartledge (*https://oreil.ly/B-vKL*).

For more advice on writing Pythonic code, check out *The Hitchhiker's Guide to Python* (*https://oreil.ly/ddReB*).

To review the benefits of Parquet files for bulk data, read "Why parquet files are my preferred API for bulk open data" by Robin Linacre (*https://oreil.ly/OmtmF*).

Extending Your Portfolio Project

Here are a few ideas to extend your SDK project:

- Finish all the endpoint functions remaining. (You didn't think I forgot, did you?)
- Publish your SDK on the PyPI Test Repository (*https://test.pypi.org*) to learn the process of deploying packages. You're more than halfway there—continue from the "Generating distribution archives" (*https://oreil.ly/IaQad*) section on the Python packaging tutorial.
- If you are creating your own portfolio project, create a minimum viable SDK for it using this chapter as a guide.

Summary

In this chapter, you learned from experts in SDK development to identify the features that make a great Python SDK. Then, you went out and coded it! SDKs like the one you developed make life much easier for data scientists and other Python-centric users of your API. While coding the SDK, you applied PEP 8 code style and used a variety of Pythonic techniques such as list comprehensions and context managers.

In Chapter 8, you will start looking at APIs from the perspective of the consumer instead of the producer. You will start by learning the skills that data scientists should know about APIs.

Using APIs in Your Data Science Project

In Part II, you will consume the APIs you built in your data science project using industry-standard libraries and specifications:

- Chapter 8 covers some topics that data scientists need to know about APIs and introduces your second portfolio project.
- In Chapter 9, you will learn to use APIs for data analytics with Jupyter Notebooks.
- In Chapter 10, you will build your API and data science skills by calling APIs in a data pipeline built with Apache Airflow.
- In Chapter 11, you will build a Streamlit data app using data from an API.

What Data Scientists Should Know About APIs

Working with APIs for data science is a necessary skill set for all data scientists.
—Nate Rosidi, KDnuggets

API expertise is critical to being an effective data scientist. But a data scientist can't become an expert in every API specialty—the field of APIs is nearly as wide as the field of data science. Thankfully, you don't need to master every API specialty if you use the *building-block approach*: mastering one or two API-related skills at a time, and stacking additional skills on top of those as your skills grow. In my experience, the best way to acquire these building blocks is through hands-on coding projects that you share with the world for fast feedback. (That's where the *hands-on* part of this book's title comes from.)

The following are some of the most useful building-block skills for data scientists.

Using a Variety of API Styles

Out in the wild, there are a few major API architectural styles that you may come across. Chapter 2 discussed why an API provider might create a REST, GraphQL, or gRPC API. As an API consumer, you need to be flexible. This section explains how you can use whatever API style is available.

The most common API style is REST or RESTful. (For simplicity, I will use the terms interchangeably in this chapter.) A REST API has multiple *endpoints*, which are combinations of HTTP verbs and URLs. For example, to read league information from a football API, you might use an HTTP GET verb and the URL *https://api.sportsworld central.com/v0/leagues*. To create a new league, you might use a POST verb with the same URL, and pass along information to it in the HTTP body.

A web browser is an easy way to send a GET request to an API. For example, to query a fantasy football API for a specific football player record with a `player_id` of 1491, you could open your web browser to a fantasy sports API at *https://api.sportsworldcen tral.com/v0/players/1491* and get a result like Figure 8-1. This is an HTTP GET.

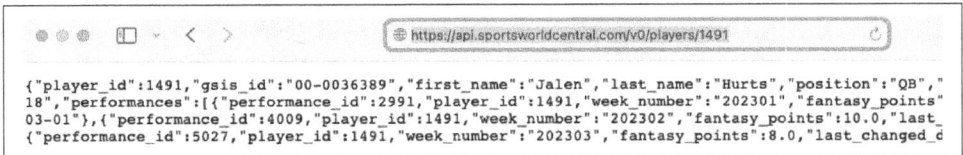

Figure 8-1. GET request of a REST API

To make the same request with Python, you can use the httpx library as follows:

```
import httpx

rest_url = "https://api.sportsworldcentral.com/v0/players/1491"

api_response = httpx.get(rest_url)
print(api_response.json())
```

This API returns data in JSON format, and it contains the standard fields that are available for this API endpoint.

Whether you called this API from the web browser or from Python, you executed an HTTP GET verb, which is used for reading information from an API. If you wanted to call a REST API to take other actions, you might use HTTP verbs such as POST, PUT, or DELETE. You will notice that REST APIs have separate endpoints for each action. For instance, this API would have separate endpoints to get a list of players or teams. The API defines the fields in the response, and they are the same for each request.

Another architectural style that is useful for data science is GraphQL, which has several differences from REST. Where REST APIs have multiple endpoints, a GraphQL API has only one. Where REST APIs return the same fields each time an endpoint is called, GraphQL allows the consumer to define what fields they want. Where REST uses an HTTP GET when reading data, GraphQL uses an HTTP POST. Since a web browser by default sends an HTTP GET, you can't directly call a GraphQL API in the browser. However, some GraphQL APIs provide a web interface that allows you to query the API.

To make a request to a GraphQL API in Python, you can send a query in a POST command using the httpx library as follows:

```
import httpx

graphql_url = "https://countries.trevorblades.com"
```

```
json_query = {'query': '''
{
  country(code: "US") {
    name
    native
    currency
    languages {
      code
      name
    }
  }
}
'''}

api_response = httpx.post(graphql_url, json=json_query)
print(api_response.json())
```

The data is returned in JSON format and contains only the fields that you requested as follows:

```
{'data': {'country': {'name': 'United States', 'native': 'United States',
'currency': 'USD,USN,USS', 'languages': [{'code': 'en', 'name': 'English'}]}}}
```

(Thanks to Trevor Blades for his sample GraphQL API and the example code (*https://github.com/trevorblades/countries*).)

The last API architectural style that I will mention is gRPC, and it is quite a bit different from REST or GraphQL. gRPC enables cross-language *remote procedure calls*, which means that your program code can call an external gRPC service like a local one. Data scientists are most likely to use gRPC when calling a machine learning model, such as a large language model.

gRPC uses a data format called *protocol buffers* instead of JSON, and it uses HTTP/2, which is a different communications protocol from the HTTP/1 protocol that is typically used by GraphQL and REST. These two differences allow gRPC to have very fast communication and support two-way data streaming instead of the request/response communication that the other two APIs use.

The Python code examples for calling a gRPC API are a bit too complicated to show in this introductory section, but a Python quickstart (*https://oreil.ly/nhwZ1*) is available if you would like to learn more about this.

HTTP Basics

Most APIs use HTTP, so it helps to learn more about it. The first tip is pretty simple: only use APIs with *HTTPS* in the URL—this means that all the API traffic will be encrypted in transit.

Two more HTTP items to understand are HTTP verbs and HTTP status codes. HTTP verbs are called *methods* by the official HTTP standards document (*https://oreil.ly/m5TTp*), which says a method "indicates the purpose for which the client has made this request and what is expected by the client as a successful result."

When you call an API in a web browser, you are using a GET method, which asks for a read-only copy of a resource or list of resources. Table 8-1 lists common HTTP verbs that are used for REST APIs.

Table 8-1. HTTP verbs and REST API usage

HTTP verb (method)	Use with APIs	Example
GET	Read a resource or list of resources.	GET *api.sportsworldcentral.com.com/players/*
POST	Create a new resource.	PUT *api.sportsworldcentral.com.com/team/*
PUT	Update an existing resource.	PUT *api.sportsworldcentral.com.com/players/1234*
DELETE	Remove an existing resource.	DELETE *api.sportsworldcentral.com.com/players/1234*

With the GET and DELETE requests in Table 8-1, the information in the URL is sufficient to perform the command, but POST and PUT would need information for creating or updating the player. This is the purpose of the HTTP message body. For APIs, the body contains JSON or XML data that the API uses to perform the action. As mentioned earlier, for GraphQL APIs, you always send a POST request, and the body of the message contains the query that you are sending to the API.

When the request is processed by the API's server, it returns an HTTP response, which has a *status code*, a numeric code that tells you if it was able to process your request. If all goes well, the response will have a status code of 200—meaning success —and the data you asked for if you were expecting any. Success doesn't always occur, and Table 8-2 lists some other status codes you may encounter when calling APIs.

Table 8-2. HTTP status codes

Status code	Typical meaning for API calls
2XX	Status codes beginning with 2 indicate success.
200 OK	The request was successful.
201 Created	A POST method successfully created a resource.
3XX	Status codes beginning with 3 indicate redirection.
301 or 308 Moved Permanently	The API address has moved permanently, so you should change your API call.
302 Moved Found	The API address redirected temporarily. Keep using the address you used.
4XX	Status codes beginning with 4 indicate client error.
400 Bad Request	Your request has an error or invalid request.
401 Unauthorized	Invalid credentials to make the API call.
404 Not Found	The resource doesn't exist or the address is wrong.

Status code	Typical meaning for API calls
5XX	Status codes beginning with 5 indicate server error.
500 Internal Server Error	Something failed unexpectedly on the server.
503 Service Unavailable	Temporary issue with service. Retry may be appropriate.

The official HTTP standards document (*https://oreil.ly/rfc_1*) gives more detailed information about the status codes.

How to Consume APIs Responsibly

The code samples earlier in this chapter show that APIs can be called easily with a few lines of code. This ease of use is probably one of the reasons that APIs have become so widespread in software development and data science.

But when using APIs for real-world analysis and applications, there are additional items you need to consider, including the following:

Follow the terms of service.
>   When using a new API, start by reading the terms of service. This will tell you up front what expectations and requirements the API providers have for you to use their API. For example, the MyFantasyLeague API terms of service (*https://oreil.ly/X3Jcf*) state that the APIs are free to use, but they can't be used to cheat in your fantasy league or harvest user data. Terms of service often list the default rate limiting that is in place, for example, saying that no user should make more than 1,000 API requests per hour. This prevents one user from swamping the service, or even the API provider's website if they are hosted on the same infrastructure. MyFantasyLeague doesn't list specific rate limits but forbids users from overloading the service (even by accident) and requests that users cache slow-changing data locally to reduce network traffic.

Handle retries gently.
>   You may want to enable an automated retry process in cases of temporary errors that may occur on the server side. To avoid overwhelming the service (and possibly getting your access disabled), consider implementing a backoff and retry process, as discussed in Chapter 7.

Handle credentials safely.
>   Most API publishers have some method of registering users for their APIs, even if access is free. This allows them to monitor how you are using their APIs and contact you about upcoming changes to their APIs. There are a variety of API authentication methods used by API providers, such as usernames, passwords, API keys, secret keys, tokens, and others. All API credentials should be stored securely and implemented in your code using a secrets manager or environment variables. Credentials should never be stored in program code or in files that will

be included in a code repository. If credentials get exposed somehow, deactivate them immediately and get new ones. Google includes additional tips in "Best practices for securely using API keys" (*https://oreil.ly/E_WRf*).

Validate inputs and outputs.
As an API consumer, you should handle data you receive from APIs carefully, and consider risks such as *SQL injection*, which is when bad actors try to send destructive SQL commands in fields where data is expected. You should also ensure that the data you send to APIs fits data types they expect.

Log and diagnose errors.
If you consume an API in a recurring data pipeline, you need to handle and log errors. Your code should handle logging in an organized fashion, to track error messages and informational messages. This will be useful to debug any issues that you encounter and be able to verify previous executions of your code.

Separation of Concerns: Using SDKs or Creating API Clients

An important principle in software development is *separation of concerns* (SoC), which means that a computer program should be broken up into chunks that perform a specific task. Consuming APIs responsibly may involve some fairly complicated logic, and code calling APIs should be separated from the rest of your code. For data scientists who like to build large projects in a single Jupyter Notebook, this may be a hard habit to adjust to. But the time it takes to implement will save you headaches in operation.

If the API provider publishes an SDK, you should use that. SDKs can provide advanced features that are created to work with a specific API. Many times, they are published on the PyPI package repository (*https://pypi.org*) and can be installed using `pip`.

For example, the Python SDK created in Chapter 7 for the SportsWorldCentral API could be published on PyPI. Figure 8-2 shows how this SDK would look on the PyPI repository.

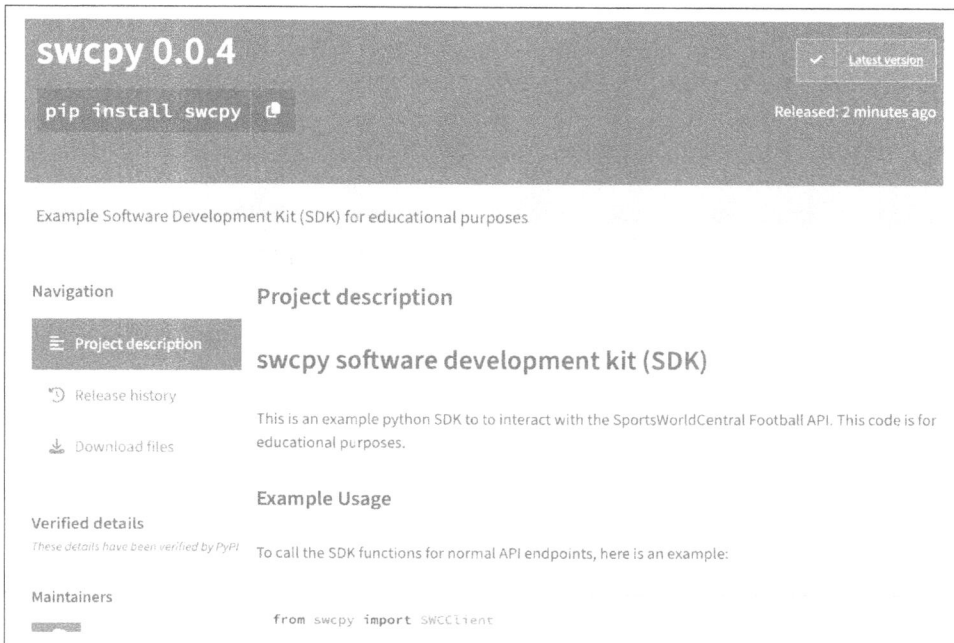

Figure 8-2. Example SDK on PyPI

To use this SDK to call the API, the following Python code would be used:

```
from pyswc import SWCClient
from pyswc import SWCConfig

config = SWCConfig()
client = SWCClient(config)
player_response = client.get_player_by_id(352)
```

The SDK is simple to call, but it includes advanced features such as backoff and retry, data validation, error handling, and logging. This is an example of the benefits that come from using an SDK when one is available.

If no SDK is available, you can separate your own API-calling code into a standalone client. You can implement the same type of functionality that an API provider's SDK would contain, but it is code you will need to maintain. In Chapter 9, you will create a Python client for the SWC API and use it in a Jupyter Notebook. You will reuse this client in Chapter 11 to call the SWC API in a Streamlit data app. This demonstrates how separating your API-calling code into a client file saves time and effort.

How to Build APIs

Data scientists increasingly need to build APIs to share the work they are doing. For instance, you may have custom metrics you have created and you would like to make them available for other data scientists to use in their work. An API is an efficient and useful way to share this data. In addition to sharing data, you may have a statistical model or machine learning model that you would like to make available. You can create an API as an *inference endpoint*, which allows users to use the model to predict outcomes based on data they submit. You will learn this in Chapter 12. Even if you are primarily a user of APIs rather than a builder, you can benefit by building a few APIs yourself to view from the other side of the desk. Part I of this book is entirely focused on building APIs—check it out if you haven't already!

How to Test APIs

Testing APIs is critical for API producers and consumers. API producers will perform testing throughout their development, deployment, and maintenance phases of hosting an API. They are responsible for ensuring that an API is reliable and that it lives up to customer expectations and any *service level agreements* (SLAs), which are formal agreements that producers make with consumers about uptime, performance, or other aspects of API service. API consumers will need to test APIs before using them in their systems to ensure that they work as intended.

Postman recommends (*https://oreil.ly/aET_m*) four major types of API testing:

Contract testing
Verifies the format and behavior of each endpoint

Unit testing
Confirms the behavior of an individual endpoint

End-to-end testing
Tests workflows that use multiple endpoints

Load testing
Verifies performance items such as the number of concurrent requests that can be processed at peak times and the response time for individual requests

The pytest library is a Python testing library that is straightforward to use. Chapters 3 and 4 show how to use it to test SQLAlchemy database code and FastAPI APIs. One Python load-testing library is Locust (*https://locust.io*). Figure 8-3 shows an example Locust load test measuring the number of requests per second that an API can process with multiple concurrent users.

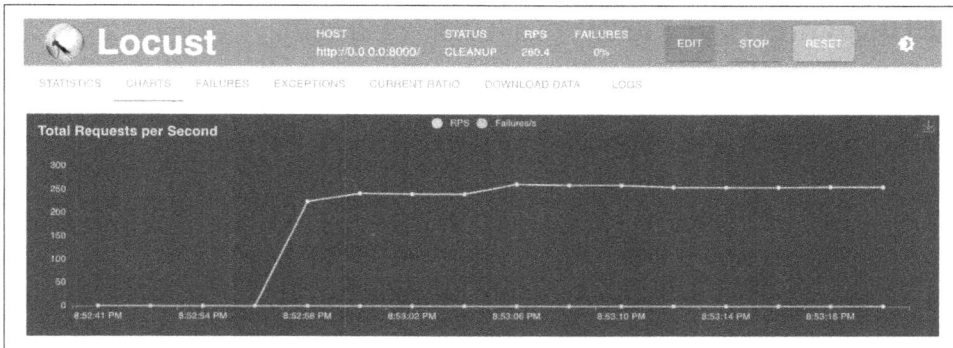

Figure 8-3. Example Locust load test report

There are additional types of testing beyond these four. The agile testing quadrants from Janet Gregory and Lisa Crispin provide a big-picture view of testing. Comprehensive testing includes technology-facing tests such as unit testing and performance testing and business-facing tests such as prototyping and usability tests. Don't forget to include your API documentation and SDKs in your testing. To learn more about the four quadrants, read "Quick Tools for Agile Testing" (*https://oreil.ly/j1ihN*).

API Deployment and Containerization

To share an API, you have to deploy it. The typical deployment model for APIs is using a cloud host, although if your users are strictly internal, you may be deploying to an on-premises server. Many cloud hosts support *containerization*, which is packaging your program code into a reusable package that can be run locally or on another server or cloud provider. Docker is the most prevalent containerization software. Chapter 6 demonstrates deploying your API to two different clouds and containerizing your API with Docker.

Using Version Control

> Version control is a way of tracking what changes have been made to a codebase, and it allows multiple people to work on the same code easily.
>
> —Catherine Nelson, *Software Engineering for Data Scientists* (O'Reilly, 2024)

This item isn't limited to working with APIs—it is a foundational skill that all data scientists will benefit from. Managing your code with version control will save hours of frustration when you're working alone and really makes things easier when you're working as a team. This part of the book continues the use of GitHub as the version control repository for your code and a place to showcase your work. You will also perform all of your development in GitHub Codespaces.

Introducing Your Part II Portfolio Project

In this part of the book, you will create a portfolio project that demonstrates your ability to create analytics and other data science products that use APIs as a source. Here is an overview of the work ahead of you:

- Chapter 9: Using APIs in data analytics products using Jupyter Notebooks
- Chapter 10: Using APIs in data pipelines using Apache Airflow
- Chapter 11: Using APIs in a Streamlit data application

Each of these tasks will enable you to showcase your API and data science skills differently.

> As you go through each chapter in this part, follow the instructions and create the code yourself. You'll learn much more by doing this than reading alone. If you run into any trouble, the files in the \complete folder are available to check your work. If you would like to complete the chapters out of order, you can also use the completed files from the previous chapters as the starting point.

Getting Started with Your GitHub Codespace

As you did in Part I, you will use GitHub Codespaces for all the code you develop. If you didn't create a GitHub account yet, do that now.

Cloning the Part II Repository

All of the Part II code examples are contained in this book's GitHub repository (*https://github.com/handsonapibook/api-book-part-two*).

To clone the repository, log in to GitHub and go to the Import Repository page (*https://github.com/new/import*). Enter the following information:

- *The URL for your source repository*: **https://github.com/handsonapibook/api-book-part-two**
- *Your username for your source code repository*: Leave blank.
- *Your access token or password for your source code repository*: Leave blank.
- *Repository name*: **analytics-project**
- *Public*: Select this so that you can share the results of the work you are doing.

Click Begin Import. The import process will begin and the message "Preparing your new repository" will be displayed. After several minutes, you will receive an email notifying you that your import has finished. Follow the link to your new cloned repository.

Launching Your GitHub Codespace

In your new repository, click the Code button and select the Codespaces tab. Click "Create codespace on main." You should see a page with the status "Setting up your codespace." Your Codespace window will be opened as the setup continues. When the setup completes, your display will look similar to Figure 8-4.

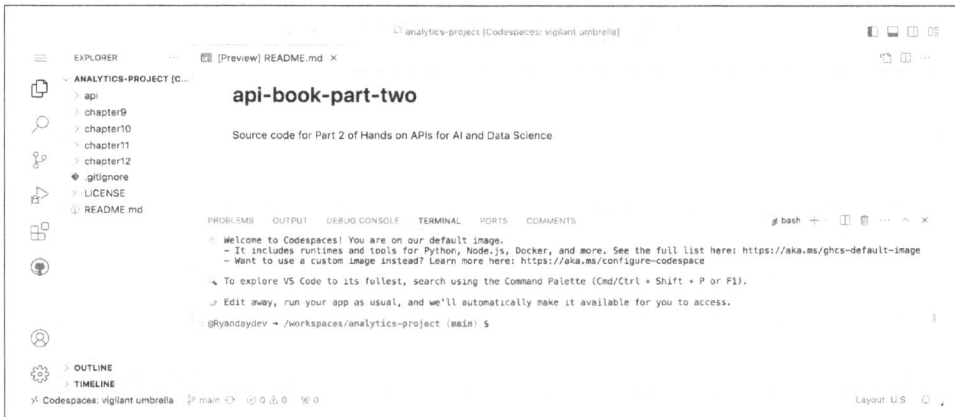

Figure 8-4. GitHub Codespace for Part II

Your Codespace is now created with the cloned repository. This is the environment you will be using for this part of the book. Open the GitHub Codespaces page (*https://github.com/codespaces*) to make a couple of updates. Scroll down the page to find this new Codespace, click the ellipsis to the right of the name, and select Rename. Enter the name **Part II Portfolio project codespace** and click Save. You should see the message "Your codespace *Part II Portfolio project codespace* has been updated." Click the ellipsis again and then click the ribbon next to "Auto-delete codespace" to turn off auto-deletion.

> To save screen real estate, I have trimmed the directory listing in the terminal prompt of the Codespace used in the examples. You can do this in your Codespace by editing the */home/code-space/.bashrc* file in VS Code. Find the `export PROMPT_DIRTRIM` statement and set it to `export PROMPT_DIRTRIM=1`. To load the values the first time, execute this terminal command: `source ~/.bashrc`.

Running the SportsWorldCentral (SWC) API Locally

As you work through the projects in Part II, you will be calling version 0.2 of the SportsWorldCentral (SWC) API, which is in the */api* folder. Version 0.2 has a few endpoints that you did not create in Part I of the book. These were created to demonstrate additional functionality in data science and AI projects. It also has additional sample data added. You will run the API in your Codespace and then call it from projects that you create in Jupyter, Airflow, and Streamlit.

Table 8-3. Updated endpoints for the SWC Fantasy Football API v0.2

Endpoint description	HTTP verb	URL	Update made
Read week list	GET	*/v0/weeks/*	New endpoint with max potential scoring
Read counts	GET	*/v0/counts/*	Added week count
Read team list	GET	*/v0/teams/*	Added weekly_scores for each team
Read league list	GET	*/v0/leagues/*	Added league size to calculate max scoring
Read individual league	GET	*/v0/leagues/{league_id}*	Added league size

In the terminal, install the required libraries in your Codespace as shown, using the *requirements.txt* file that is provided:

```
.../analytics-project (main) $ cd api
.../api (main) $ pip3 install -r requirements.txt
```

Verify that the FastAPI CLI was loaded so that you can run your API from the command line as shown:

```
.../api (main) $ pip3 show fastapi-cli
Name: fastapi-cli
Version: 0.0.4
Summary: Run and manage FastAPI apps from the command line with FastAPI CLI.
[results truncated for space]
```

Now launch the API from the command line as shown:

```
.../api (main) $ fastapi run main.py
```

You will see several messages from the FastAPI CLI, ending with the following:

```
INFO:     Started server process [19192]
INFO:     Waiting for application startup.
INFO:     Application startup complete.
INFO:     Uvicorn running on http://0.0.0.0:8000 (Press CTRL+C to quit)
```

You will see a dialog stating "Your application running on port 8000 is available," as shown in Figure 8-5. Click Make Public.

Port		Forwarded Address		Running Process
8000	⬦ ×	https://studio... 🔗 ⊕ 🗗		/usr/local/python/3.12.1,
Add Port			Open in Browser	

Figure 8-5. Makeing the API public

The API is now running in Codespaces with a public port. To view the API in the browser, click Ports in the terminal and hover your cursor over Port 8000, as shown in Figure 8-6.

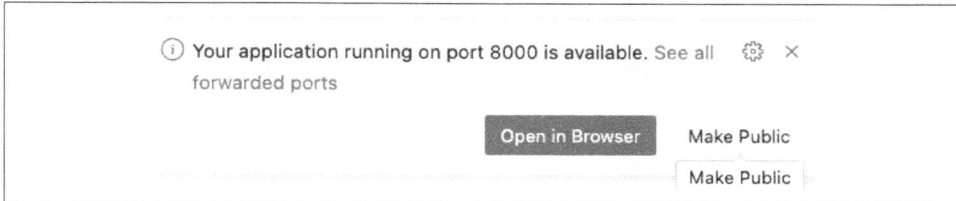

ⓘ Your application running on port 8000 is available. See all ⚙ ×
forwarded ports

Open in Browser   Make Public

Make Public

Figure 8-6. Open API on a public address

Click the globe icon. The browser will show a base URL that ends in *app.github.dev* that contains the response from your API running on Codespaces. You should see the following health check message in your web browser:

```
{"message":"API health check successful"}
```

Your API is running publicly in the cloud. Copy this base URL. You will use it in later chapters to connect to your API.

> Later chapters in Part II will instruct you to launch your API in Codespaces. Follow these instructions to run the API. The API will need to be restarted each time you reconnect to Codespaces.

Additional Resources

To learn more about the features of REST, GraphQL, and gRPC, read Chapter 2.

To learn about creating data APIs, read Part I (and create a portfolio project).

To learn how to package API client code and create SDKs, read Chapter 7.

Summary

This chapter covered some of the basic topics that data scientists need to know about APIs. You learned about major API architectural styles, including REST, GraphQL, and gRPC. You learned the basics of HTTP verbs and status codes. You learned about using APIs responsibly and the benefits of an SDK. Finally, you saw the value of learning to build, test, and deploy your APIs.

In Chapter 9, you will start to dig into the details of these topics as you use APIs in data analytics.

Using APIs for Data Analytics

Your eyes see the game much better than the numbers. But the numbers see all the games.
    —Dean Oliver, sports statistician

The sports world loves all forms of *data analytics*—charts, graphs, and statistics that describe the results of events or predict what will happen next. When a sports fan views those data analytics, they probably never consider what data source was used to create them. In many cases, the data source is an API. In this chapter, you will learn best practices for consuming APIs and creating data analytics products using Jupyter Notebooks, a popular tool used by data scientists.

Custom Metrics for Sports Analytics

One of the most celebrated forms of analytics is the *custom metric*, a calculation that summarizes complicated behavior, ability, and outcomes as a number. Every sport has metrics that players, coaches, managers, and fans pay attention to. Baseball has the longest history with metrics, from the historical *earned run average* (ERA) to the modern *weighted runs created plus* (wRC+) and *wins against replacement* (WAR). Soccer fans and professionals alike focus on the *expected goals* (xG), a method of defining quality shots that has motivated a variety of secret-sauce models (*https://oreil.ly/ HHSL0*). The NBA uses the *player efficiency rating* (PER) to measure a basketball player's all-around value.

Some of the most interesting work in custom metrics is happening in football, where the NFL has sponsored an annual analytics contest called the Big Data Bowl (*https:// oreil.ly/5h1RE*), where data students and professionals research and propose new metrics for prizes, job prospects, and the excitement of seeing their work included in TV broadcasts. It has a special track devoted to creating new custom metrics, such as *converted tackle opportunity* and *path analysis via swarm-tackle accuracy* (PASTA). Some have made their way into the broadcast booth, and more are on the way.

Every custom metric requires a few components to succeed:

Question
What question is it trying to solve? It should be general enough to have broad application but specific enough to add new knowledge to the sport. For example, the KenPom ranking (*https://oreil.ly/2MIk3*) answers the question: which college basketball teams deserve to make the NCAA tournament field?

Theory
By choosing specific numbers to measure and weighing them against each other, you make a value judgment. You are proposing subcomponents that matter to answer a question.

Valid approach
Do the underlying calculations support the purpose of the metric?

Data source
Can you get data to calculate the metric at a reasonable frequency? If data isn't available, your approach and the supporting theory may have to be adjusted out of practicality.

Name
This is the fun part. The more interesting the name, the more impact it can have.

Using APIs as Data Sources for Fantasy Custom Metrics

Fantasy sports enthusiasts love analytics and metrics too. Fantasy league websites provide a lot of stats and charts, but commissioners and managers sometimes want to create their own. Since the fantasy league data is updated frequently, an automated process is key to making the calculations repeatable and consistent. To gather this data in software code, two primary choices are available: APIs or *web scraping*. Web scraping involves using program code to read the HTML from a website page and extract the data. The technique is powerful but brittle: every time the website structure changes, the web scraper code stops working and has to be modified and tested again. But if the league website maintains an API, the data is available even when changes to the layout or structure of the web page occur.

A web search turns up various examples of fantasy managers who share their custom metrics for others to use. The most feature-rich metric I have found is the Leeger Python library (*https://oreil.ly/leegpy*), which is maintained by software engineer Joey Greco. It is an open source project that generates custom metrics from six fantasy football websites. Appropriate for this chapter, Leeger uses APIs heavily.

You read some of Joey's advice about SDKs in Chapter 7. I talked to him about how he uses APIs to create metrics with Leeger.

API Perspectives: Joey Greco on APIs and Custom Metrics

When did you first start using APIs from the fantasy websites?

Early in 2022, I realized that copying the fantasy scores for my leagues each week into my own database was tedious and prone to error. One particularly frustrating evening (the Tuesday after I went 0-4 in my fantasy leagues), I had reached my limit, and I looked into what APIs were available for the fantasy sites I was using at that time (ESPN and MyFantasyLeague).

What motivated you to create the Leeger app and the fantasy SDKs for the league host websites?

I really like keeping stats for all of my fantasy leagues. Over the years, I slowly evolved from calculating stats in a paper notebook, to using Excel, to manually entering them in a database, to writing code that can retrieve data from my fantasy sites and calculate them for me. This led me to create a Python SDK that anyone could use to do the same thing (pull data automatically from their fantasy site and have stats calculated for them each week). While developing Leeger, I realized that some fantasy sites either didn't have a Python SDK or had one that was old/not maintained. I figured while I'm at it I might as well make them myself, so that's what I did!

What are some of your favorite custom metrics that you have developed for your leagues?

I think my favorite to date is AWAL, which stands for "Adjusted Wins Against the League." It's a simple metric that shows how many wins a team should have had if schedule was not a factor. It looks at any given week and assigns each team a score based on how they ranked in that week's scoring. You can look at this AWAL over the course of the season and see clear trends of over- and underperformance. Other metrics I've created, like Team Score, Team Success, and Team Luck, are a lot of fun. I'm always thinking about new ways to look at fantasy statistics and extract more meaning from them. It's part of what makes it so much fun for me!

The Leeger app that Joey created contains dozens of custom metrics that are calculated from data extracted from league website APIs. They have entertaining names, such as Adjusted Team Luck and Smart Wins. Here is the calculation behind the AWAL stat that Greco mentioned:

```
AWAL = (teams_outscored_in_week * (1/possible_opponents_in_week)) +
(teams_tied_in_week * (0.5/L))
```

Creating a Custom Metric: The Shark League Score

Now it's your turn to create a custom metric: the Shark League Score. Figure 9-1 shows the high-level architecture of the project you will create in this chapter.

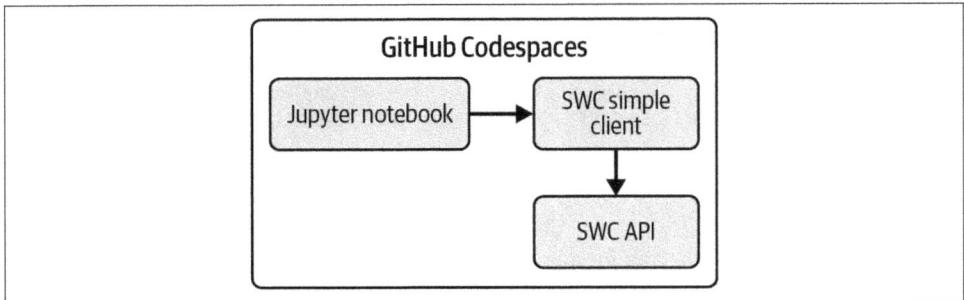

Figure 9-1. High-level architecture

The question it seeks to answer is: "How tough is our fantasy league?" A real fantasy shark is a manager who knows their stuff: they are prepared for the draft, they scour the waiver wire for the best players, and they always seem to start the right lineup. A league filled with owners like this is a Shark League, and making the playoffs in this type of league is a badge of honor. Winning a title in a Shark League—that is a victory to savor for a lifetime.

The theory of the Shark League Score is based on a few attributes of efficient managers. First, Shark Leagues should be balanced from top to bottom—there should be no easy weeks on the schedule. Second, the league as a whole should be picking up the best players and getting them in rosters to score points. One supporting theory is that fantasy regular season stats (usually weeks 1 through 14) are more appropriate than fantasy playoffs (week 15 and beyond) because all teams are playing.

There are a few additional items that could arguably be included but are not supported by the SportsWorldCentral API. For instance, a Shark League should rarely or never have an empty roster spot in the weekly starters. Another sign of efficient owners is picking up the "hot free agents" before they have their big weeks—that would be especially tricky to measure.

As you go through the examples in this chapter, you will develop the Shark League Score, as well as two supporting metrics: the League Balance Score and League Juice Score. It's time to get started!

Software Used in This Chapter

Table 9-1 lists a few of the software components you will begin using in this chapter.

Table 9-1. Key tools or services used in this chapter

Software name	Purpose
backoff	Python library for adding backoff and retry to web calls
httpx	Python library for making web calls
Jupyter Notebooks	Interactive data science environment
pandas	Data analysis and formatting library

httpx

The Python library you will use for calling APIs is httpx. This library is very similar to the popular requests library but also supports asynchronous API calls. For more information about httpx, read Chapter 4.

You will use version 27.x of httpx to stay consistent with the version used in Part I.

Jupyter Notebooks

Jupyter Notebooks support a unique way to perform data science. This mode is called *interactive computing*, and it allows you to mix code cells, Markdown comment cells, and results. Cells flow from the top of a notebook to the bottom, and variables and libraries that have been run previously are available to cells that run later. (If you run cells only sequentially, that would mean the cells below, but you can run cells out of order.) In addition to this interactive mode, Jupyter Notebooks also provide Markdown cells that run between code cells, which allows you to create richly formatted documents that interweave code, results, and documentation. The notebook style of programming is heavily used by data scientists, who value its ability to store the results of prior work along with the code that went into creating it.

Jupyter Notebooks are supported out of the box with the default VS Code installation in your GitHub Codespace. You will run notebooks directly in VS Code by creating a file with the *.ipynb* extension. Project Jupyter has many more options and features beyond what you will learn in this chapter; you can learn more about them at the Jupyter project home page (*https://jupyter.org*).

You will use the default version of Jupyter that comes installed in Codespaces.

pandas

For data scientists using Python, one of the most trusted libraries is pandas. For data scientists, it is one of the first imports you include in almost any Python program. The pandas library provides Python a data type called a `DataFrame`, which is a two-dimensional structure with rows and columns. This spreadsheet-like format is a natural way of viewing data in rows and columns.

The library also provides many methods for data manipulation, filtering, and formatting. As mentioned in Chapter 1 of this book, data scientists spend more than one-third of their time preparing and cleansing data. The pandas library is a powerful tool for this type of work. The official pandas user guide (*https://oreil.ly/X4vXa*) is a great reference for this library.

You will use the default version of pandas that comes installed in GitHub Codespaces.

Installing the New Libraries in Your Codespace

To install the libraries you need for this chapter, create a file named *chapter9/requirements.txt*:

```
.../analytics-project (main) $ cd chapter9
.../chapter9 (main) $ touch requirements.txt
```

Update *chapter9/requirements.txt* with the following contents:

```
logging ❶
httpx>=0.27.0
backoff>=2.2.1 ❷
```

❶ This is the standard Python logging module.

❷ This library provides backoff and retry functionality for API calls. See Chapter 7 for more details.

Execute the following command to install the new libraries in your Codespace:

```
.../chapter9 (main) $ pip3 install -r requirements.txt
```

You should see a message that states that these libraries were successfully installed.

Launching Your API in Codespaces

To access your API data, you will need to launch v0.2 of your API in the terminal. For instructions, read "Running the SportsWorldCentral (SWC) API Locally" on page 176. Copy the URL of your API from the browser address bar to use as the base URL in this chapter.

Creating an API Client File

You will create a standalone Python file to make all calls to your API. By maintaining this file separate from the Jupyter Notebook, you keep special API-related logic in one place and make it available to multiple notebooks.

You will use the backoff library to implement exponential backoff and retry with jitter, which makes your API calls more reliable without overwhelming the source API. You use the HTTPX client in a context manager style. Chapter 7 uses these techniques and a few more to create a full-featured Python SDK. Using SDKs when available is also helpful for Jupyter Notebooks, but for this chapter, you will add most of the features yourself.

> Version 0.2 of the API has several new endpoints that you didn't create in Part I. To explore the format of the new endpoints, access the interactive API docs using Swagger UI at the */docs* endpoint on your API.

Create a new Python file in the terminal as shown:

```
.../analytics-project (main) $ cd chapter9
.../chapter9 (main) $ mkdir notebooks
.../chapter9 (main) $ touch notebooks/swc_simple_client.py
```

Update *swc_simple_client.py* with the following code:

```
import backoff
import logging
import httpx

HEALTH_CHECK_ENDPOINT = "/" ❶
LIST_LEAGUES_ENDPOINT = "/v0/leagues/"
LIST_PLAYERS_ENDPOINT = "/v0/players/"
LIST_PERFORMANCES_ENDPOINT = "/v0/performances/"
LIST_TEAMS_ENDPOINT = "/v0/teams/"
LIST_WEEKS_ENDPOINT = "/v0/weeks/"
GET_COUNTS_ENDPOINT = "/v0/counts/"

logger = logging.getLogger(__name__) ❷

@backoff.on_exception( ❸
    wait_gen=backoff.expo,
    exception=(httpx.RequestError, httpx.HTTPStatusError),
    max_time=5,
    jitter=backoff.random_jitter
)
def call_api_endpoint(
    base_url: str,
    api_endpoint: str,
```

```
        api_params: dict = None  ❹
) -> httpx.Response:

    try:
        with httpx.Client(base_url=base_url) as client:  ❺
            logger.debug(f"base_url: {base_url}, api_endpoint: {api_endpoint}")
            response = client.get(api_endpoint, params=api_params)
            response.raise_for_status()
            logger.debug(f"Response JSON: {response.json()}")  ❻
            return response
    except httpx.HTTPStatusError as e:
        logger.error(f"HTTP status error occurred: {e.response.text}")  ❼
return httpx.Response(status_code=e.response.status_code,
                      content=b"API error")  ❽
    except httpx.RequestError as e:
        logger.error(f"Request error occurred: {str(e)}")
        return httpx.Response(status_code=500, content=b"Network error")
    except Exception as e:
        logger.error(f"Unexpected error occurred: {str(e)}")
        return httpx.Response(status_code=500, content=b"Unexpected error")
```

❶ The URL endpoints are set as variables that can be used when calling the client.

❷ This statement gets a reference to the log file.

❸ This decorator adds backoff and retry functionality to the call_api_endpoint
    function. For more information about the settings, see the backoff documenta-
    tion (*https://oreil.ly/c_4yV*).

❹ This allows you to pass in parameters to the API.

❺ This statement uses the HTTPX client in a resource manager style, which makes
    the API call and then cleans up resources when it finishes.

❻ This statement logs the data in the API response for debugging.

❼ If errors occur, they are logged with an ERROR type.

❽ In case of error, the client returns an httpx.response object with the error code
    and message.

Creating Your Jupyter Notebook

To get started, run the following commands in the Terminal window at the bottom of
the screen to create the new directory and the Jupyter Notebook you will be using in
this chapter:

```
.../analytics-project (main) $ cd chapter9
.../chapter9 (main) $ mkdir notebooks
.../notebooks (main) $ touch notebooks/shark_league_notebook.ipynb
```

If you open the *chapter9/notebooks* folder in the Explorer on the left, you should see *shark_league_notebook.ipynb*. Click to open it. As shown in Figure 9-2, you will see a blank cell. A Jupyter Notebook is made up of cells like these that you can fill with software code to run commands or Markdown-formatted text to provide context and explanation for the code.

In the top-right of the file you will see Select Kernel, as shown in Figure 9-2.

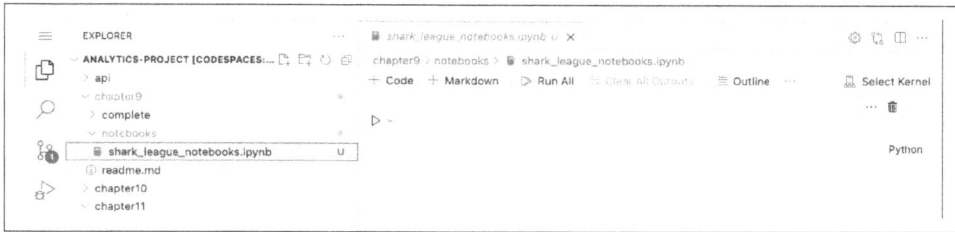

Figure 9-2. New notebook file

Click Select Kernel. Codespaces should prompt you to "Install/Enable suggested extensions Python + Jupyter" as shown in Figure 9-3. Select "Install/Enable suggested extensions Python + Jupyter." Click Install in the additional pop-up window, if prompted.

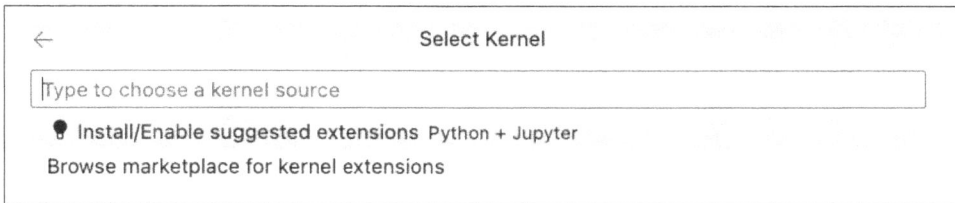

Figure 9-3. Install/enable extensions

After the installation completes, the title of the window will change to Select Another Kernel and you will see the choice Python Environments. Select Python Environments.

The title of the window will change to "Select a Python Environment." One Python version should be listed with a star next to it and the label Recommended—select this Python version.

Adding General Configuration to Your Notebook

The beginning of your notebook will contain general configuration and setup. Hover your cursor above the empty Python cell and click +Markdown to create a new Markdown cell. Enter the following title in the Markdown cell:

```
# Shark League Score
## Import Python libraries
```

Run this cell by clicking the play icon on the left of the cell or by pressing Shift-Enter. You should see your message formatted as a title.

Hover your cursor below this cell and click "+Code" to create a new Python cell. Enter the following code in the Python cell:

```
import pandas as pd
import logging
import swc_simple_client as swc  ❶
```

❶   This references the Python file you created named *swc_simple_client.py*.

Placing all the imports at the top of your notebook helps keep track of the libraries you are using. These imports will work for all the cells in this Jupyter Notebook.

> Jupyter Notebooks are executed top to bottom. Although the output of Jupyter Notebooks is saved between sessions, the variables and imported libraries are not. When you start a coding session, select Execute Above Cells to rerun all the cells above the one you are using.

As mentioned in Chapter 8, logging is an important component of working with APIs. You will configure a log file named *shark_notebook.log* to store the logging messages that are generated in your notebook. These log files are excluded by your repository's *.gitignore* file, so they will not be committed to your repository, which is a good practice.

Add another Markdown cell with the following text:

```
## Configure logging
```

Add and run a Python code cell with the following:

```
for handler in logging.root.handlers[:]:  ❶
    logging.root.removeHandler(handler)

logging.basicConfig(
    filename='shark_notebook.log',
    level=logging.INFO,  ❷
)
```

❶ This statement removes any existing logging handlers configured by Codespaces.

❷ This sets the logging level to record in the log. Review Table 7-1 for more details about Python logging.

The next code cell will contain shared variables, which are good to add after the import statements. In this notebook, you set a reusable variable for the base URL of the API and created string constants with the API endpoints. These two steps make the purpose of API calls clearer and help avoid manual typing errors. These will be available to all the cells in the notebook. Add another Markdown cell with the following text:

```
## Setup notebook variables
```

Add and run a Python code cell with the following:

```
base_url = "[insert your API base URL]" ❶
```

❶ Inside the quotations, you should put the base URL of the API running locally on Codespaces, without the backslash, for example, *https://fluffy-lemur-12345-8000.app.github.dev*.

Working with Your API Data

The next several cells use the imported `call_api_endpoint()` function to call the API and get an `httpx.Response` object. Then, they extract the API data using `Response.json()` and store it in a pandas DataFrame.

Add a Markdown cell with the following text:

```
## Get Max Scores
### Use endpoint: LIST_WEEKS_ENDPOINT
```

This code retrieves the maximum scores for SportsWorldCentral leagues based on custom scoring types, which will be used for the League Juice Score. Highly custom values like these are examples of data that is best retrieved from the website's API. Although you could potentially estimate the total possible points using public NFL scoring data, it would take a lot of effort and would likely be slightly different from the final totals the website calculates. Since the website makes this available, you can easily get an exact match to the league totals.

From this point, the contents of the code cell and the output will be displayed together. The code will be on top and the output will follow the OUTPUT statement.

Add and run a Python code cell with the following:

```
week_api_response = swc.call_api_endpoint(base_url,swc.LIST_WEEKS_ENDPOINT)
weeks_df = pd.DataFrame(week_api_response.json())  ❶
weeks_df['year'] = weeks_df['week_number'].str.slice(0, 4).astype(int)  ❷

max_totals_grouped_df = weeks_df.groupby('year').agg(  ❸
    ppr_12_max_points=('ppr_12_max_points', 'sum'),
    half_ppr_8_max_points=('half_ppr_8_max_points', 'sum'))

display(max_totals_grouped_df)

OUTPUT:
year ppr_12_max_points  half_ppr_8_max_points
2023            21048.0                14800.0
```

❶ This extracts the data in a dictionary format and creates a `DataFrame`.

❷ This uses the pandas `str.slice` method to get the year substring from the `week_number`.

❸ This uses the pandas `groupby().agg()` method to group the data by year and calculate the maximum points for two of the league types.

Next, you'll retrieve the league information. Add a Markdown cell with this text:

```
## Get League Scoring Type
### Use Endpoint: LIST_LEAGUES_ENDPOINT
```

Add and run a Python code cell with the following:

```
league_api_response = swc.call_api_endpoint(base_url,swc.LIST_LEAGUES_ENDPOINT)
leagues_df = pd.DataFrame(league_api_response.json())
leagues_df = leagues_df.drop(columns=['teams','last_changed_date'])  ❶
display(leagues_df)

OUTPUT:
    league_id                   league_name scoring_type  league_size
0        5001  Pigskin Prodigal Fantasy League        PPR           12
1        5002       Recurring Champions League   Half-PPR            8
2        5003                    AHAHFZZFFFL   Half-PPR            8
3        5004   Gridiron Gurus Fantasy League        PPR           12
4        5005                Best League Ever        PPR           12
...
```

❶ This statement uses the pandas `drop` method to exclude columns from the Data-Frame.

Next, you will retrieve the total scoring for each league to compare to the max potential. Add a Markdown cell with the following text:

```
## Get Regular Season Scoring Totals - By Team
### Use Endpoint: LIST_TEAMS_ENDPOINT
```

This section of code includes two techniques that are useful when using pandas to process API data. The SportsWorldCentral API returns JSON data with several nested dictionaries in the `weekly_scores` element, but you want to have one row per week. You also want to get multiple columns out of the nested column. You will use the pandas `json_normalize()` function to accomplish these tasks.

This section also introduces the pandas `groupby().sum()` method, which is similar to the SQL `GROUP BY` statement. You will use this to total up all of the weekly scoring values and give a total for the entire fantasy regular season. (The NFL plays 18 weeks, but SportsWorldCentral considers weeks 1 through 14 the fantasy regular season, and the rest are the playoffs.)

Add and run a Python code cell with the following:

```
team_api_response = swc.call_api_endpoint(base_url,swc.LIST_TEAMS_ENDPOINT)

teams_parsed_df = pd.json_normalize(team_api_response.json(), 'weekly_scores',
                        ['league_id', 'team_id', 'team_name'])  ❶

teams_parsed_df['year'] = (teams_parsed_df['week_number']
                        .str.slice(0, 4).astype(int))  ❷
teams_parsed_df['week'] = (teams_parsed_df['week_number']
                        .str.slice(4, 6).astype(int))

#get only regular season teams
teams_regular_season_df = teams_parsed_df.query('week <= 14')

#get team season totals
team_totals_df = teams_regular_season_df.groupby(
    ['league_id', 'team_id', 'year'], as_index = False
    )['fantasy_points'].sum()  ❸

team_totals_df.head()
```

❶ The pandas `json_normalize()` function breaks the nested JSON data into multiple rows and extracts new columns.

❷ The pandas `str.slice()` method is used to extract the `year` and `week` values from the `week_number` field.

❸ This `groupby` statement sums the fantasy points by league, team, and year.

Calculating the League Balance Score

With the data loading and formatting out of the way, you can calculate your first metric: the League Balance Score. The intuition behind this metric is that a high-quality league is a balanced league. Instead of being dominated by one or two top teams, the league has a balance of teams that are all competitive.

A balanced league has less variability among the team's regular season totals. One way to measure for variability is to calculate the *standard deviation* of the values. However, leagues with high scoring systems tend to have a higher variability, making it difficult to compare leagues with different scoring systems. To adjust for this, you'll use the *coefficient of variation* (CV) of each league's regular season total scores. This takes the *standard deviation* of the league's totals and divides it by the *mean* of the totals.

This gives you a measure of relative variability between the values that adjusts for the overall scoring system. The reason CV works for this situation is that it is *dimensionless*, which means it can be compared across values of difference sizes—scoring systems in this case. Here is the exact formula you will use:

$$LeagueBalanceScore = 100 - stdev(LeagueRegularSeasonTotal)/mean(LeagueRegularSeasonTotal) * 100$$

A CV is lower if it varies less, but you want a metric where a higher number is better. You also want it to be comparable to the League Juice Score, which has a max value of 100. To accomplish this, you multiply it by 100 and subtract it from 100 to give a number similar in scale to the League Juice Score, and so that a larger number is better, also matching the League Juice Score.

Add another Markdown cell, with the following text:

```
## League Balance Score
### Using the Coefficient of Variation (CV) of league regular season totals
```

This section also uses a `lambda` command to execute Python code during the aggregation. You will use this in several other locations.

Add and run a Python code cell with the following:

```
league_stats_df = team_totals_df.groupby(['league_id','year']).agg(
    league_points_sum=('fantasy_points', 'sum'),       ❶
    league_points_mean=('fantasy_points', 'mean'),
    league_points_stdev=('fantasy_points', 'std'),     ❷
  league_balance_score=('fantasy_points',
                        lambda x: (100 -(x.std() / x.mean()) * 100))  ❸
).reset_index()

display(league_stats_df.sort_values(by='league_balance_score',
                                    ascending=False))  ❹
```

❶ This sum value will be used in the next metric, but this is a convenient place to calculate it.

❷ This uses the pandas built-in `GroupBy.std` calculation for the standard deviation of the league totals.

❸ This uses `lambda` to execute a calculation on the aggregated values to calculate the league score and scale it to match the League Juice Score.

❹ The pandas `sort_values` method does not change the underlying structure; it only sorts during the display.

The output of this cell is shown in Figure 9-4.

	league_id	year	league_points_sum	league_points_mean	league_points_stdev	league_balance_score
1	5002	2023	9126.0	1140.750000	58.801725	94.845345
0	5001	2023	14052.0	1171.000000	128.872735	88.994643
3	5004	2023	13752.0	1146.000000	148.203177	87.067786
4	5005	2023	13313.0	1109.416667	145.786555	86.859170
2	5003	2023	8599.0	1074.875000	157.835753	85.315897

Figure 9-4. League Balance Score

Calculating the League Juice Score

The second metric you will calculate is the League Juice Score, which is the percentage of potential points that the league scored for the season. This represents how much juice the league owners squeezed out of the orange, that is, how many potential points were in the starting lineups. In fantasy football, it doesn't do much good for the week's top scorers to be sitting on someone's bench. In a high-quality league, managers are setting starting lineups that get the most from their teams.

One wrinkle to the calculation is that max points differ by the size of the league and the scoring type. Before you can calculate the score, you need to merge three DataFrames that you already prepared into a single DataFrame:

league_stats_df
    Contains the total points scored by each team and year

max_totals_grouped_df
    Contains the custom max point totals for the regular season

leagues_df
    Contains the scoring type and league size of each league, to match against custom max points

The exact formula you will use is the following:

$$LeagueJuiceScore = 100 * (LeagueTotalPoints) / MaxPotentialPoints)$$

Add another Markdown cell with the following text:

```
## League Juice Score
### Compare league scoring to max potential scoring
```

Add and run a Python code cell with the following:

```
league_stats_with_league_max_df = (league_stats_df[
    ['league_id','year', 'league_points_sum','league_balance_score']]
            .merge(max_totals_grouped_df,left_on = 'year', right_on='year'))❶

combined_metrics_df = (leagues_df[
    ['league_id','league_name','scoring_type', 'league_size']]❷
    .merge(league_stats_with_league_max_df,
        left_on = 'league_id', right_on = 'league_id'))

combined_metrics_df['league_juice_score'] = combined_metrics_df.apply(❸
    lambda row: (
        100 * (row['league_points_sum'] / row['ppr_12_max_points'])❹
        if (row['scoring_type'] == 'PPR' and row['league_size'] == 12)
        else (
            100 * (row['league_points_sum'] / row['half_ppr_8_max_points'])
            if (row['scoring_type'] == 'Half-PPR' and row['league_size'] == 8)
            else None
        )
    ),
    axis=1
)

combined_metrics_df = (combined_metrics_df.drop(
    columns=['scoring_type','league_size','league_points_sum'
            ,'ppr_12_max_points','half_ppr_8_max_points',] )
)
display(combined_metrics_df)
```

❶ The first step is to combine `league_stats_df` with `max_totals_grouped_df`.

❷ Next, you combine `leagues_df` with the output of the previous step.

❸ This section uses the `apply()` method to execute a `lambda` function against each row to calculate the custom `league_juice+score` value.

❹ You multiply the ratio by 100 to make it a percentage, and scale it to match the League Balance Score.

The output of this cell is shown in Figure 9-5.

	league_id	league_name	year	league_balance_score	league_juice_score
0	5001	Pigskin Prodigal Fantasy League	2023	88.994643	84.063173
1	5002	Recurring Champions League	2023	94.845345	74.171001
2	5003	AHAHFZZFFFL	2023	85.315897	69.887841
3	5004	Gridiron Gurus Fantasy League	2023	87.067786	82.268485
4	5005	Best League Ever	2023	86.859170	79.642259

Figure 9-5. League Juice Score

Creating the Shark League Score

After your API data has been manipulated, munged, and merged, you are ready for the payoff: building your Shark League Score. Based on your decision to weight the League Juice Score double, the final formula you will use is:

$$SharkLeagueScore = 2 * (LeagueJuiceScore) + LeagueBalanceScore$$

Add another Markdown cell, with the following text:

```
# Create Shark League Score
## Shark League Score = (2 * League Juice Score) + League Balance Score
```

Add and run a Python code cell with the following:

```
combined_metrics_df['shark_league_score'] = combined_metrics_df.apply(
    lambda league: (2 * league['league_juice_score']) +
    league['league_balance_score'],
    axis=1
)
display(combined_metrics_df.sort_values(by='shark_league_score',
ascending=False))
```

The output of this cell is shown in Figure 9-6.

	league_id	league_name	year	league_balance_score	league_juice_score	shark_league_score
0	5001	Pigskin Prodigal Fantasy League	2023	88.994643	84.063173	257.120989
3	5004	Gridiron Gurus Fantasy League	2023	87.067786	82.268485	251.604756
4	5005	Best League Ever	2023	86.859170	79.642259	246.143688
1	5002	Recurring Champions League	2023	94.845345	74.171001	243.187348
2	5003	AHAHFZZFFFL	2023	85.315897	69.887841	225.091580

Figure 9-6. Shark League Score

Congratulations, you have calculated the Shark League Score based on customized data from the SportsWorldCentral API.

Additional Resources

For a discussion of the value of statistics in basketball, read "NBA Insider: Is It Numbers or Talent? Sorting Fact, Fiction in NBA Stats Wave" (*https://oreil.ly/DqSG-*).

To continue building your knowledge of pandas, I recommend *Python for Data Analysis: Data Wrangling with pandas, NumPy, and Jupyter* (O'Reilly, 2022). It was written by Wes McKinney, the creator of pandas.

To learn more detailed charts that can be created with football data from nfl_data_py, I recommend the book *Football Analytics with Python & R: Learning Data Science Through the Lens of Sports* Eager and Erickson (O'Reilly, 2023).

For more tips on formatting Markdown for your notebook, read the Markdown Guide (*https://oreil.ly/_7OgB*).

To learn more about the coefficient of variation in other domains, read "Coefficient of Variation: Meaning and How to Use It" (*https://oreil.ly/yO4zn*).

If you'd like to explore more uses of data analytics in fantasy football using Python, Nathan Braun has courses and books available at *https://fantasycoding.com*.

Summary

In this chapter, you learned about creating custom metrics using API data. You calculated the Shark League Score and learned how to use pandas and Jupyter Notebooks along the way.

In Chapter 10, you will build your API and data science skills by calling APIs in a data pipeline built with Apache Airflow.

Using APIs in Data Pipelines

In their simplest form, pipelines may extract only data from one source such as a REST API and load to a destination such as a SQL table in a data warehouse. In practice, however, pipelines typically consist of multiple steps ... before delivering data to its final destination.

—James Densmore *Data Pipelines Pocket Reference* (O'Reilly, 2021)

In Chapter 9, you used a Jupyter Notebook to query APIs and create data analytics. Querying directly in a notebook is useful for exploratory data analysis, but it requires you to keep querying the API over and over again. When data teams create analytics products for production, they implement scheduled processes to keep an up-to-date copy of source data in the format they need. These structured processes are called *data pipelines* because source data flows into the pipeline and is prepared and stored to create data products. Other common terms for these processes are *Extract, Transform, Load (ETL)* or *Extract, Load, Transform (ELT)*, depending on the technical details of how they are implemented. *Data engineer* is the specialized role that focuses on the development and operation of data pipelines, but in many organizations, data scientists, data analysts, and infrastructure engineers also perform this work.

In this chapter, you will create a data pipeline to read SportsWorldCentral fantasy football player data using Apache Airflow, a popular open source tool for managing data pipelines using Python.

Types of Data Sources for Data Pipelines

The potential data sources for data pipelines are almost endless. Here are a few examples:

APIs

> REST APIs are the focus of this book, and they are an important data source for data pipelines. They are better suited for incremental updates than full loads, because sending the full contents of a data source may require many network calls. Other API styles such as GraphQL and SOAP are also common.

Bulk files

> Large datasets are often shared in some type of bulk file that can be downloaded and processed. This is an efficient way to process a very large data source. The file format of these may vary, but CSV and Parquet are popular formats for data science applications.

Streaming data and message queues

> For near-real-time updates of data, streaming sources such as Apache Kafka or AWS Kinesis provide continuous feeds of updates.

Message queues

> Message queue software such as RabbitMQ or AWS SQS provides asynchronous messaging, which allows transactions to be published in a holding location and picked up later by a subscriber.

Direct database connections

> A connection to the source database allows a consumer to get data in its original format. These are more common for sharing data inside organizations than to outside consumers.

You will be creating a pipeline that uses REST APIs and bulk files in this chapter.

Planning Your Data Pipeline

Your goal is to read SportsWorldCentral data and store it in a local database that you can keep up to date. This allows you to create analytics products such as reports and dashboards. For this scenario, you'll assume that the API does not allow full downloads of the data, so you will need to use a bulk file for the initial load.

After that initial load, you want to get a daily update of any new records or records that have been updated. These changed records are commonly referred to as *delta* or *deltas*, using the mathematical term for "change." By processing only the changed records, the update process will run more quickly and use fewer resources (and spend less money).

Figure 10-1 displays the data pipeline you are planning.

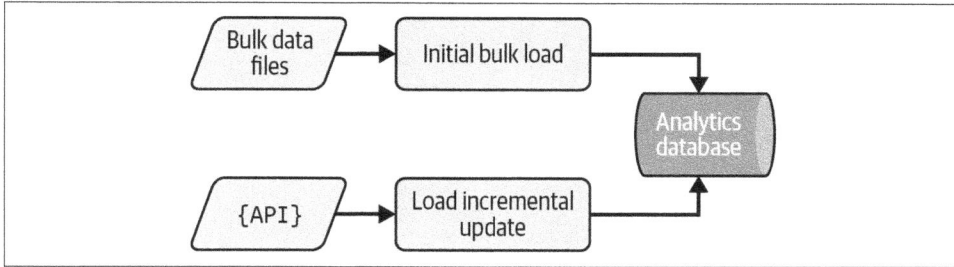

Figure 10-1. Plan for your data pipeline

The pipeline includes two sources: bulk data files and an API. The rounded boxes represent two ETL processes and they both will update the analytics database, a local database that is used to create analytics products like dashboards and reports.

Orchestrating the Data Pipeline with Apache Airflow

> *Airflow is best thought of as a spider in a web: it sits in the middle of your data processes and coordinates work happening across the different (distributed) systems.*
>
> —Julian de Ruiter and Bas Harenslak, *Data Pipelines with Apache Airflow* (Manning, 2021)

Running multiple data processing work streams in production gets complicated quickly. Scheduling, error handling, and restarting failed processes require significant planning and design. These tasks are called *orchestration*, and this is what Apache Airflow is used for. As the number of data pipelines grows, you will benefit from using orchestration software instead of coding all of these tasks yourself. Airflow is a full-featured open source engine that uses Python for its configuration, and it handles many of the recurring tasks involved in data pipelines.

Airflow has some specialized terminology that is not used in other data science programming. Astronomer's Airflow glossary (*https://oreil.ly/IjTM4*) is a complete source for these, but I will share some of the most important ones with you.

Airflow uses terminology from mathematical graph theory. In graph theory, a *node* is a process and an *edge* is a flow between nodes. Using this terminology, a *directed acyclic graph* (DAG) is a top-level process that contains steps proceeding in one direction without any loops or recursive logic.

Figure 10-2 shows how nodes and edges relate to each other in a DAG.

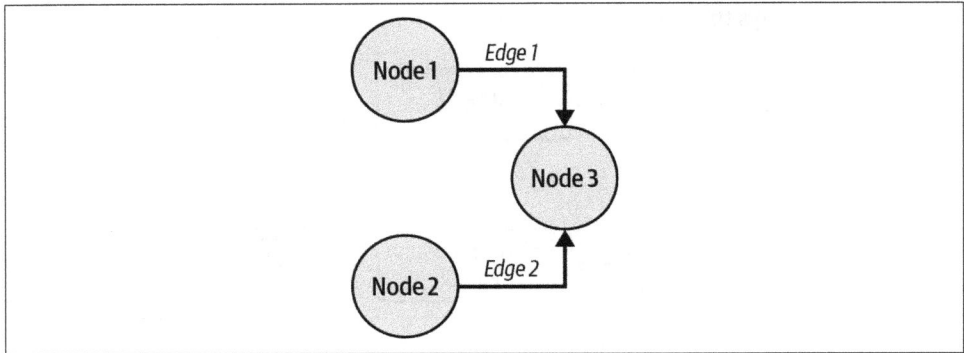

Figure 10-2. Directed acyclic graph

You will create one Python file for each DAG. Each of the steps in a DAG is called a *task*, the basic unit of execution in Airflow. Each task will be displayed as a single box on the graph diagram of a DAG.

An *operator* is a predefined template for a task. In this chapter, you will use an `Http Operator` to call your API and a `PythonOperator` to update your analytics database. Airflow has built-in operators to interact with databases, S3 buckets, and several other functions. Dozens more are available from the community and are listed in the Airflow Operators and Hooks Reference (*https://oreil.ly/8k6mr*).

The last thing you will learn to use is an *XCom*, which stands for *cross-communications*. XComs are used to pass information and data between tasks.

Installing Apache Airflow in GitHub Codespaces

Figure 10-3 shows the high-level architecture of the project you will create in this chapter.

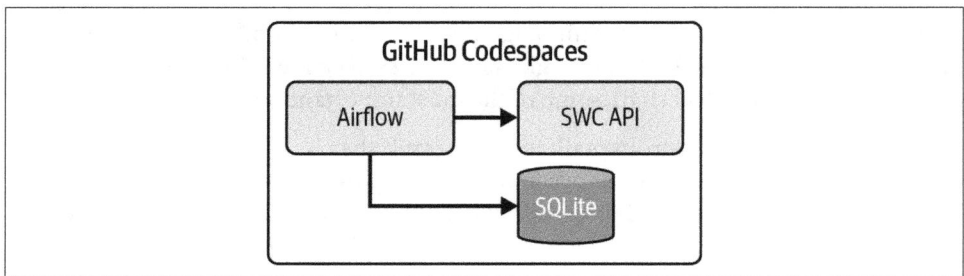

Figure 10-3. Architecture of the Airflow project

You will be working with the Part II GitHub Codespace that you created in "Getting Started with Your GitHub Codespace" on page 174. If you haven't created your Part II Codespace yet, you can complete that section now.

Before launching the Codespace, change the machine type to a four-core machine by clicking the ellipsis next to the Codespace and then clicking "Change machine type." This is necessary because Airflow runs multiple services at once.

You will be installing Airflow in the Codespace and performing that basic configuration that allows you to create the data pipeline from the diagram. (This will be a non-production setup for demonstration purposes. Before using Airflow in production, additional setup would be required.)

Airflow can be installed using Docker or pip. You will be using the Docker version. You will follow the instructions from "Running Airflow in Docker" (*https://oreil.ly/ORZKy*), with a few customizations.

To begin, create an *airflow* directory in the *chapter10* folder of your Codespace and change to that directory:

```
.../chapter10 (main) $ mkdir airflow
.../chapter10 (main) $ cd airflow
.../airflow (main) $
```

Next, use the curl command to retrieve a copy of the *docker-compose.yaml* file that is used to run the Docker version of Airflow. Get this from the official Airflow website (*https://airflow.apache.org*), and specify the version. This chapter demonstrates with version 2.9.3, but you can follow the latest stable version listed in the Airflow documentation (*https://oreil.ly/QTlk_*):

```
.../airflow (main) $ curl -LfO \
'https://airflow.apache.org/docs/apache-airflow/2.10.0/docker-compose.yaml'
  % Total    % Received % Xferd  Average Speed   Time    Time     Time  Current
                                 Dload  Upload   Total   Spent    Left  Speed
100 11342  100 11342    0     0   410k      0 --:--:-- --:--:-- --:--:--  410k
```

The file *docker-compose.yaml* contains instructions for the images to download from Docker Hub (*https://oreil.ly/q7y53*) along with environment options for configuring the software in your environment.

Open *docker_compose.yaml* and take a look at the volumes: section:

```
volumes:
  - ${AIRFLOW_PROJ_DIR:-.}/dags:/opt/airflow/dags
  - ${AIRFLOW_PROJ_DIR:-.}/logs:/opt/airflow/logs
  - ${AIRFLOW_PROJ_DIR:-.}/config:/opt/airflow/config
  - ${AIRFLOW_PROJ_DIR:-.}/plugins:/opt/airflow/plugins
```

This section creates Docker *volumes*, which are virtual drives available inside the Docker containers that are mapped to files in your Codespace storage. They are relative to the Airflow project directory, which will be *airflow* in your Codespace. For example, *airflow/dags* in your Codespace will be referenced as */opt/airflow/dags* to the Airflow application running in Docker. (This will be important when you create connections later in this chapter.)

Create the directories that are mapped to those volumes and then configure an environment variable for the Airflow user ID:

```
.../airflow (main) $ mkdir -p ./dags ./logs ./plugins ./config
.../airflow (main) $ echo -e "AIRFLOW_UID=$(id -u)" > .env
```

Create *docker-compose.override.yaml*:

```
.../airflow (main) $ touch docker-compose.override.yaml
```

You will use this file to override some of the standard configuration settings from the *docker-compose.yaml* file you downloaded. Using an override file allows you to keep the *docker-compose.yaml* file exactly like you downloaded it and put all of your customizations together, which makes troubleshooting easier. It also allows you to update *docker-compose.yaml* with a new version when Airflow is upgraded. Update *docker-compose.override.yaml* with the following contents:

```
#these are overrides to the default docker compose
x-airflow-common:
  &airflow-common
  environment:
    &airflow-common-env
    AIRFLOW__CORE__LOAD_EXAMPLES: 'false' ❶

services:
  airflow-webserver:
    <<: *airflow-common
    command: webserver
    environment:
      <<: *airflow-common-env
      AIRFLOW__WEBSERVER__ENABLE_PROXY_FIX: 'True' ❷
  airflow-scheduler:
    <<: *airflow-common
    command: scheduler
    environment:
      <<: *airflow-common-env
      AIRFLOW__SCHEDULER__MIN_FILE_PROCESS_INTERVAL: '30' ❸
```

❶ This setting will hide the built-in Airflow examples so that they are not distracting in this chapter.

❷ This setting will allow you to use the Airflow web interface in Codespaces.

❸ This setting tells Airflow to look for changes to your code more frequently while you are developing.

Now you are ready to initialize the Docker environment using *docker-compose.yaml* and *docker-compose.override.yaml* with the docker compose up airflow-init command. This command will download the Airflow software and provision user IDs and other configuration details. Execute the following command:

```
.../airflow (main) $ docker compose up airflow-init
[+] Running 44/3
  redis Pulled
  postgres Pulled
  airflow-init Pulled
...
airflow-init-1  | 2.10.0
airflow-init-1 exited with code 0
```

This command will run for several minutes, with many commands executed. If the output ends with "exited with code 0" it was successful. Your environment has been initialized, and you don't need to execute this command again.

You are ready to run Airflow. To launch the Airflow web interface, execute the following command:

```
.../airflow (main) $ docker compose up -d
+] Running 7/7
 ✓ Container airflow-postgres-1           Healthy
 ✓ Container airflow-redis-1              Healthy
 ✓ Container airflow-airflow-init-1        Exited
 ✓ Container airflow-airflow-webserver-1   Started
 ✓ Container airflow-airflow-triggerer-1   Started
 ✓ Container airflow-airflow-scheduler-1   Started
 ✓ Container airflow-airflow-worker-1      Started
```

Although you will see a pop-up window to launch the web UI, I have found that sometimes the web UI takes a few minutes to prepare, so don't click OK. Instead, wait a couple of minutes and then select the Ports tab in your Codespace. You will see the forwarded address of the web interface. Click the globe icon to open the UI in the browser, as shown in Figure 10-4.

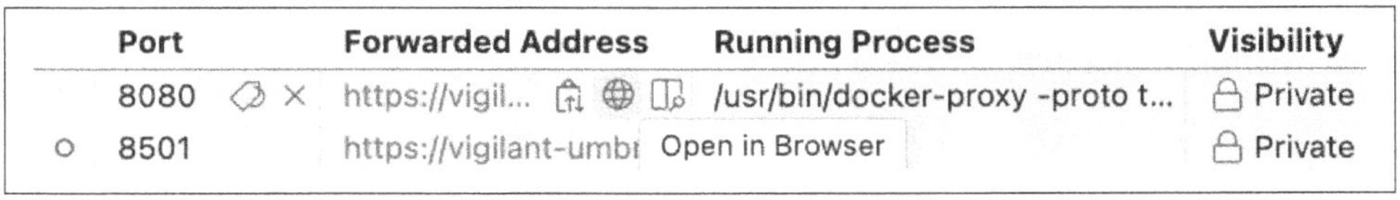

	Port		Forwarded Address	Running Process	Visibility
	8080	⟳ ×	https://vigil... 🔗 🌐 🗐	/usr/bin/docker-proxy –proto t...	🔒 Private
○	8501		https://vigilant-umbr	Open in Browser	🔒 Private

Figure 10-4. Open Airflow web interface

You will see the login page. Enter a username of **airflow** and password of **airflow** and click "Sign in." (These starter credentials are used for demonstration only.) You will see the web interface of the Airflow application running in your Codespace, as shown in Figure 10-5. When you begin, there are no DAGs listed. You will learn more about the capabilities of Airflow as you create DAGs to complete your data pipeline requirements.

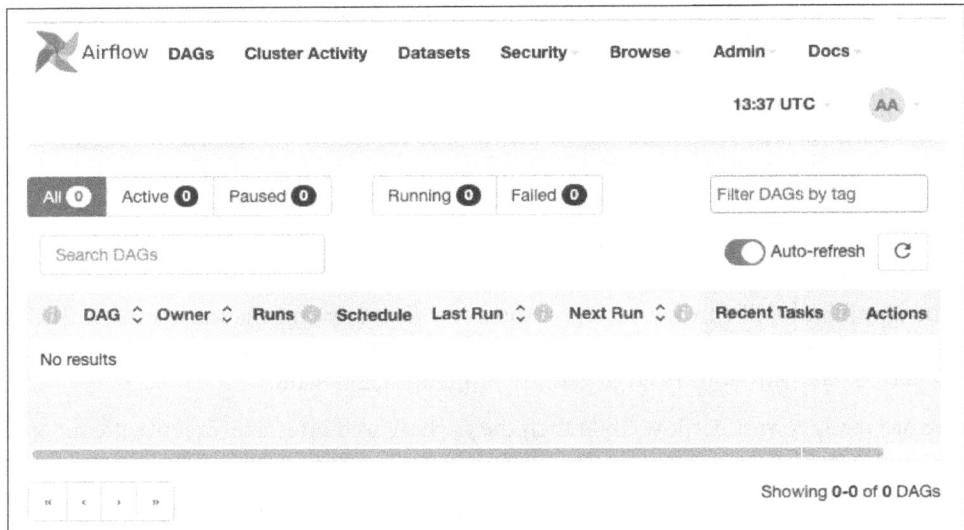

Figure 10-5. Airflow home page

Creating Your Local Analytics Database

Your data pipeline will be used to insert and update player records into a local database. This is a common data science pattern: updating a database from source data and then creating models, metrics, and reports from the database. Change the directory to *dags* and create a database and the `player` table as follows:

```
.../airflow (main) $ cd dags
.../dags (main) $ sqlite3 analytics_database.db
SQLite version 3.45.3 2024-04-15 13:34:05
Enter ".help" for usage hints.
sqlite> CREATE TABLE player (
    player_id INTEGER PRIMARY KEY,
    gsis_id TEXT,
    first_name TEXT,
    last_name TEXT,
    position TEXT,
    last_changed_date DATE
);
sqlite> .exit
```

Launching Your API in Codespaces

Your Airflow pipeline needs a running copy of the SportsWorldCentral API to gather updates. Follow the directions in "Running the SportsWorldCentral (SWC) API Locally" on page 176 to get your API running in a separate terminal window of Codespaces, and copy the base URL from the browser bar. You will configure Airflow to reference the base URL from that API in the next section.

Configuring Airflow Connections

Airflow *connections* allow you to store information about data sources and targets in the server instead of in your code. This is useful for maintaining separate Airflow environments for development, testing, and production. You will create connections for your API and your analytics database.

In the Airflow UI, select Admin > Connections. Click the plus sign to add a new connection record. Now you will use the *volume* mappings that you viewed earlier in the *docker-compose.yaml* file. Use the following values:

- *Connection ID*: `analytics_database`
- *Connection Type*: Sqlite
- *Description*: `Database to store local analytics data.`
- *Schema*: `/opt/airflow/dags/analytics_database.db`

Leave the rest of the values empty, and click Save.

Next, add the connection for the API connection:

- *Connection ID*: `sportsworldcentral_url`
- *Connection Type*: HTTP
- *Description*: `URL for calling the SportsWorldCentral API.`
- *Host*: Enter the base URL of the API running in Codespaces.

Leave the rest of the values empty, and click Save. You should see two connections listed, as shown in Figure 10-6.

	Conn Id ⇅	Conn Type ⇅	Description ⇅	Host ⇅
☐ ✎ 🗑	analytics_database	sqlite	Database to store local analytics data.	
☐ ✎ 🗑	sportsworldcentral_url	http	URL for calling the SportsWorldCentral API.	https://example-url.app.github.dev/

Figure 10-6. Configured Airflow connections

Creating Your First DAG

Figure 10-7 displays an implementation of your pipeline with Airflow, using two DAGs. The *bulk_player_file_load.py* DAG would perform an initial load of the analytics database from a bulk file, which was provided in Part I of this book. That file is available in the *chapter10/complete* folder of your repository, but this chapter does not walk through it due to space constraints.

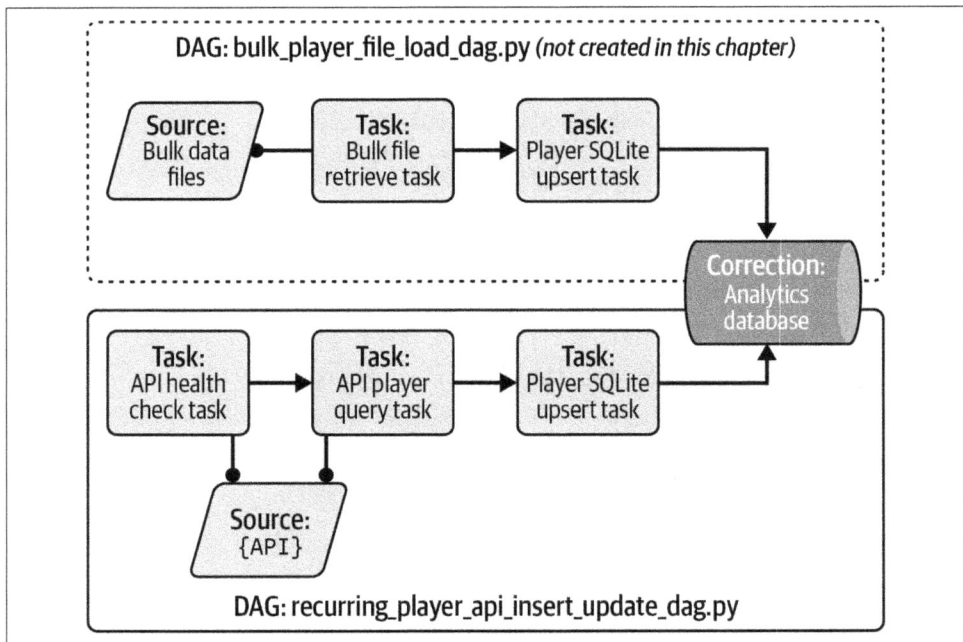

Figure 10-7. Airflow components of your data pipeline

Create the DAG that uses API data, *recurring_player_api_insert_update_dag.py*. This DAG performs incremental updates of your database, using the SportsWorldCentral API. Change the directory to *dags* and create the *recurring_player_api_insert_update_dag.py* file:

```
.../airflow (main) $ cd dags
.../dags (main) $  touch recurring_player_api_insert_update_dag.py
```

Add the following contents to the *recurring_player_api_insert_update_dag.py* file:

```
import datetime
import logging
from airflow.decorators import dag ❶
from airflow.providers.http.operators.http import HttpOperator ❷
from airflow.operators.python import PythonOperator
from shared_functions import upsert_player_data ❸

def health_check_response(response): ❹
    logging.info(f"Response status code: {response.status_code}")
    logging.info(f"Response body: {response.text}")
    return response.status_code == 200 and response.json() == {
        "message": "API health check successful"
    }

def insert_update_player_data(**context): ❺

    player_json = context["ti"].xcom_pull(task_ids="api_player_query") ❻

    if player_json:
        upsert_player_data(player_json) ❼
    else:
        logging.warning("No player data found.")

@dag(schedule_interval=None) ❽
def recurring_player_api_insert_update_dag():

    api_health_check_task = HttpOperator( ❾
        task_id="check_api_health_check_endpoint",
        http_conn_id="sportsworldcentral_url",
        endpoint="/",
        method="GET",
        headers={"Content-Type": "application/json"},
        response_check=health_check_response,
    )

    temp_min_last_change_date = "2024-04-01" ❿

    api_player_query_task = HttpOperator( ⓫
        task_id="api_player_query",
        http_conn_id="sportsworldcentral_url",
        endpoint=(
            f"/v0/players/?skip=0&limit=100000&minimum_last_changed_date="
```

```
            f"{temp_min_last_change_date}"
        ),
        method="GET",
        headers={"Content-Type": "application/json"},
    )

    player_sqlite_upsert_task = PythonOperator(  ⓬
        task_id="player_sqlite_upsert",
        python_callable=insert_update_player_data,
        provide_context=True,
    )

    # Run order of tasks
    api_health_check_task >> api_player_query_task >> player_sqlite_upsert_task⓭

# Instantiate the DAG
dag_instance = recurring_player_api_insert_update_dag()  ⓮
```

❶ This import allows you to define the DAG using a @dag decorator.

❷ These two imports allow you to use predefined operators in your tasks.

❸ This is an import of a separate Python file with a function that is shared between two DAGs.

❹ This is the code to verify the response of api_health_check_task defined below. This is the first task, and it allows the DAG to verify the status of the API before proceeding with other tasks.

❺ This defines a function that will be called by a task.

❻ This line of code uses XCom to retrieve data from the second task.

❼ Here it passes the data from XCom to the shared upsert_player_data function, which is defined in a separate file.

❽ This is the main DAG definition. It uses the @dag decorator to define the Python function as a DAG. The tasks are defined within this method.

❾ The first task uses an HttpOperator template to call the API's health check endpoint. It adds a response_check method to check the API's status before continuing.

❿ The minimum last changed date is hardcoded in this example. In production, an Airflow template variable (*https://oreil.ly/pFHaG*) could be used to get the last day's updates.

⓫ The second task uses an `HttpOperator` to call the API's player endpoint with a query parameter to restrict the records that are returned.

⓬ The third task is a `PythonOperator` that calls the `insert_update_player_data` function.

⓭ This statement sets the dependency of the tasks using bitshift operators.

⓮ The final statement is required to instantiate the DAG that is defined by the `@dag` decorator.

Take a moment to compare this code to Figure 10-7. The key parts of the DAG file are toward the bottom of this file: the `@dag` decorator defines the main DAG wrapper. Inside the DAG are three tasks: two that use `HttpOperators` to connect to the API and one that uses a `PythonOperator` to connect to the SQLite database.

The statement `api_health_check_task` >> `api_player_query_task` >> `player_sqlite_upsert_task` sets the dependency between the tasks using a right-shift operator, >>. These tasks have a very simple sequential dependency, but Airflow is capable of implementing very intricate dependencies between tasks. For more information about this capability, read Astronomer's "Manage task and task group dependencies in Airflow" (*https://oreil.ly/PTa4M*).

Coding a Shared Function

Although the sources of the two DAGs are different, they both perform an *upsert* on the analytics database, which means that if a source record already exists in the database, the code updates it, otherwise it inserts a new record. Because this task is shared between the two DAGs, you will create a separate Python file with a shared function. Create the *shared_functions.py* file:

```
.../dags (main) $ touch shared_functions.py
```

Add the following contents to the *shared_functions.py* file:

```
import logging
import json
from airflow.hooks.base import BaseHook

def upsert_player_data(player_json):
    import sqlite3                              ❶
    import pandas as pd

# Fetch the connection object
    database_conn_id = 'analytics_database'
    connection = BaseHook.get_connection(database_conn_id)  ❷
```

```
        sqlite_db_path = connection.schema

    if player_json:

        player_data = json.loads(player_json)

        # Use a context manager for the SQLite connection
        with sqlite3.connect(sqlite_db_path) as conn:
            cursor = conn.cursor()                           ❸

            # Insert each player record into the 'player' table
            for player in player_data:
                try:
                    cursor.execute(""" ❹
                        INSERT INTO player (
                            player_id, gsis_id, first_name, last_name,
                            position, last_changed_date
                        )
                        VALUES (?, ?, ?, ?, ?, ?)
                        ON CONFLICT(player_id) DO UPDATE ❺
                        SET
                            gsis_id = excluded.gsis_id,
                            first_name = excluded.first_name,
                            last_name = excluded.last_name,
                            position = excluded.position,
                            last_changed_date = excluded.last_changed_date
                    """, (
                        player['player_id'], player['gsis_id'],
                        player['first_name'],
                        player['last_name'],
                        player['position'],
                        player['last_changed_date']
                    ))
                except Exception as e:
                    logging.error(
                        f"Failed to insert player {player['player_id']}: {e}")
                    raise

    else:
        logging.warning("No player data found.")
        raise ValueError(
            "No player data found. Task failed due to missing data.")
```

❶ These two import statements are placed inside the Python method. This is because Airflow frequently parses DAG code and will reload imported libraries that are at the top of the Python file.

❷ This statement uses an Airflow hook to retrieve the connection that you defined in the Airflow user interface.

❸ This uses a database cursor to execute SQL queries on your analytics database.

❹ This statement uses the database cursor to execute a parameterized SQL query.

❺ This SQL statement provides the upsert capability, which updates a record if it already exists or inserts it if not.

This function receives the data from the API as a parameter and then loads data into the SQLite database using the Airflow connection that you defined in the user interface. This is a parameterized SQL query, in which the input data is referenced with VALUES (?, ?, ?, ?, ?, ?). This is an important measure to protect against SQL injection attacks, which could occur if a malicious actor inserted code into the source data's fields, where your process was expecting data.

Running Your DAG

Before you run the DAG, check that your API is up and running. Navigate back to the Airflow UI and you will see your DAG listed, as shown in Figure 10-8. The user interface has too many features to cover in this chapter, but you can read about the user interface at "UI / Screenshots" (*https://oreil.ly/DfOSC*).

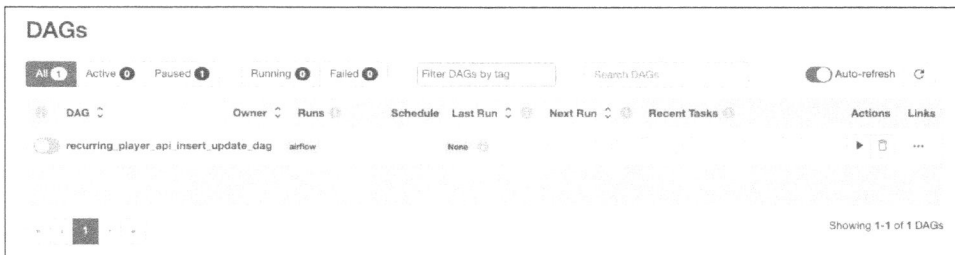

Figure 10-8. DAG listed on the Airflow home page

Click `recurring_player_api_insert_update_dag`, and then Graph. You will see the sequence of Airflow tasks using the `task_id` names that you assigned in your code, as shown in Figure 10-9.

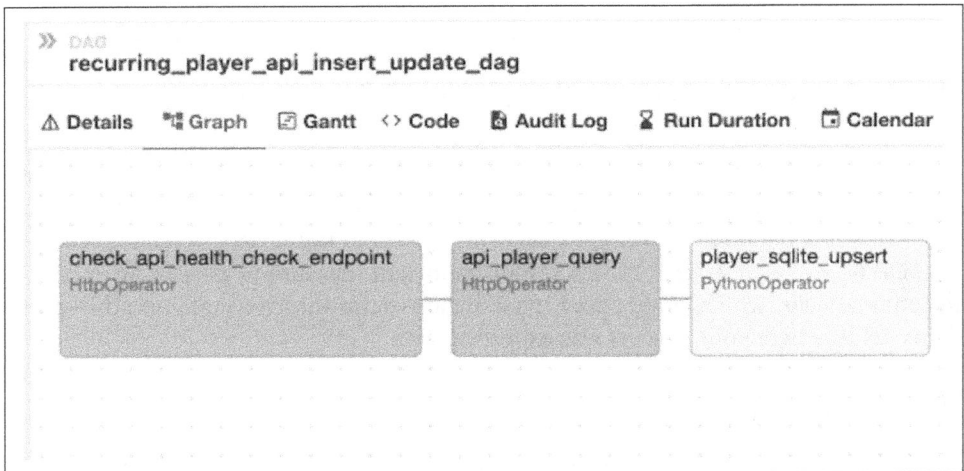

Figure 10-9. Graph view of the first DAG

Click the Trigger DAG button, which has a triangle icon to your DAG. If everything is configured correctly with your code and connections, each of the tasks in your DAG should complete with a green box in a minute or so. Click the first box, labeled `check_api_health_check_endpoint`. Your view should look similar to Figure 10-10. If you encounter an error, click the task that has the error, and click Logs to diagnose the issue.

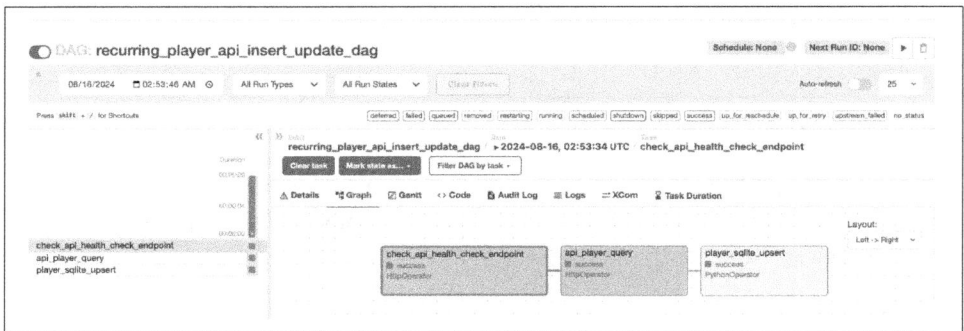

Figure 10-10. Successful DAG run

To confirm that your analytics database was successfully upserted, go back to the terminal and open the database with SQLite. Query the Player table, to confirm that 1,018 player records are present in the table. These are the records retrieved from your API:

```
.../dags (main) $ sqlite3 analytics_database.db
SQLite version 3.45.3 2024-04-15 13:34:05
Enter ".help" for usage hints.
sqlite> select count(*) from player;
1018
```

Congratulations! You created a data pipeline that updates your database with records from an API!

Extending Your Portfolio Project

Here are a few ways to extend the project you created in this chapter:

- Update *recurring_player_api_insert_update_dag.py* to pass the minimum last changed date using the scheduling variables built into Airflow (*https://oreil.ly/YViYQ*).
- The *bulk_player_file_load_dag.py* DAG is in the *chapter10/complete* directory. Create the Airflow HTTP connection mentioned in the code, and get this DAG to work.
- Add DAGs to process all of the other API endpoints.

Summary

In this chapter, you learned how to create a data pipeline calling APIs to maintain current data for analytics products. You installed and configured Apache Airflow, and you created a DAG with multiple tasks to update your database from an API.

In Chapter 11, you will create a Streamlit data app using data from an API.

Using APIs in Streamlit Data Apps

A simple app today is better than an over-designed app three months late.
—Thiago Teixeira, cofounder of Streamlit

If you want to demonstrate your data science skills to other people, it's hard to beat a *data app*—a web application displaying your data science models, graphs, charts, and spreadsheets. Sending a recruiter or potential client a link to a data app you created is a sure way to get their attention.

In this chapter, you will build a data app with Streamlit, an open source library that helps you create colorful and interactive web apps using Python. Streamlit handles all the complexity of the web interface so that you can focus on the data backend code, such as data files, pandas DataFrames, and APIs.

Streamlit apps can be deployed in a limited fashion for free on the Streamlit Community Cloud or paid web hosting platforms.

Engaging Users with Interactive Visualizations

Two main types of analytics products are *tabular reports* and *visualizations*. Tabular reports present rows and columns of data in a spreadsheet format, which provides a detailed view of data. Visualizations replace tabular data with charts, graphs, maps, and other images that give context and color to the data. Visualizations can communicate ideas that are hard to see in a spreadsheet.

Streamlit supports both tabular data and visualizations in a very interactive way. It provides built-in *widgets*, which are application controls such as sliders and select boxes that allow the user to filter and interact with the data in real time. For a fun example, take a look at Figure 11-1, which shows a data app I built to track the championship titles in one of my long-running fantasy leagues.

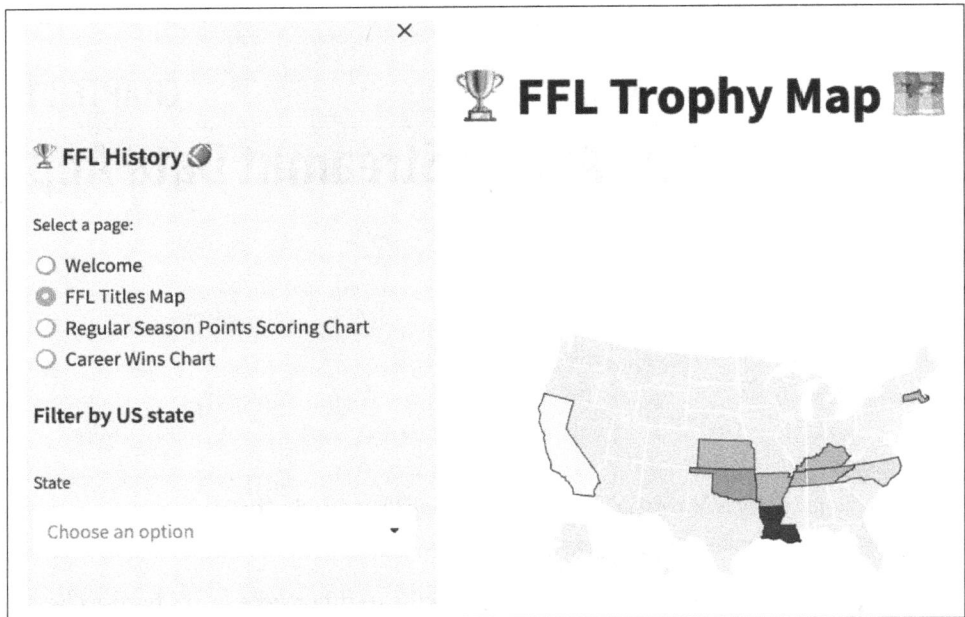

Figure 11-1. Fantasy league data app

The built-in widgets allow a data scientist to focus on providing accurate and useful data, while Streamlit components make a great-looking app. Streamlit designs its product to have simple defaults that get users up and running quickly; this is Streamlit's design principle of promoting forward progress (*https://oreil.ly/strlit*).

Software Used in This Chapter

This chapter builds on the tools you have learned in previous chapters, especially pandas. Table 11-1 displays the new tools you will use.

Table 11-1. Key tools or services used in this chapter

Software name	Purpose
nfl_data_py	NFL data library
Streamlit	Web-based analytics software

nfl_data_py

To perform data analytics, you often want to add data from multiple sources to make your analytics products more informative or give them wider context. For football data, nflverse (*https://oreil.ly/nfLV*) is a rich set of open source libraries and data files containing NFL data. The Python library for interacting with the data is nfl_data_py (*https://oreil.ly/2U7P8*).

You will use the most current version of nfl_data_py that is available.

Streamlit

Streamlit is an open source Python library that you can download using `pip3`. By including the library, a single Python file can generate an app using the command `streamlit run your_script.py`. The library is easy to learn using the Streamlit getting started guide (*https://oreil.ly/-qxBe*).

Installing Streamlit and nfl_data_py

To install the libraries you need for this chapter, create a file named *chapter11/requirements.txt*:

```
.../analytics-project (main) $ cd chapter11
.../chapter11 (main) $ touch requirements.txt
```

Update *chapter11/requirements.txt* with the following contents:

```
streamlit>=1.38.0
httpx>=0.27.0
nfl_data_py
matplotlib
backoff>=2.2.1
```

Execute the following command to install the new libraries in your Codespace:

```
.../chapter11 (main) $ pip3 install -r requirements.txt
```

You should see a message that states that these libraries were successfully installed.

Launching Your API in Codespaces

Figure 11-2 shows the high-level architecture of the project you will create in this chapter.

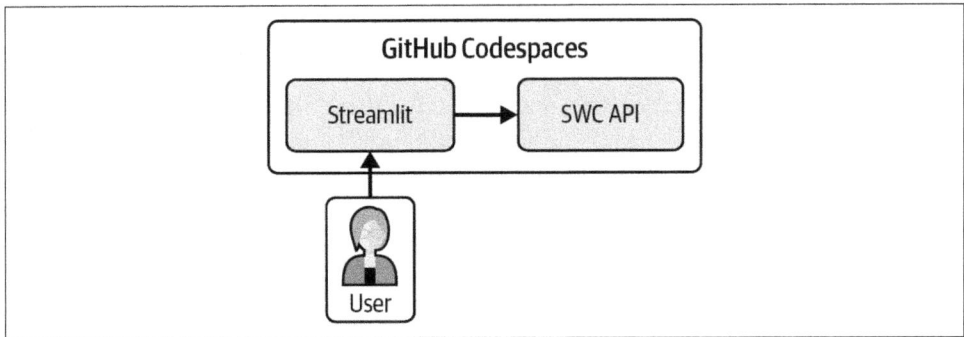

Figure 11-2. Architecture of the Streamlit project

To access your API data, you will need to launch v0.2 of your API in the terminal. For instructions, follow "Running the SportsWorldCentral (SWC) API Locally" on page 176. When the API is running, copy the URL of your API from the browser address bar to use as the base URL in this chapter.

> When your API or Streamlit app is running in the terminal and you want to stop it, press Ctrl-C.

Reusing the Chapter 9 API Client File

In Chapter 9, you created a standalone Python client file to make the calls to your API, while adding logging, error handling, and backoff-retry functionality. You will reuse this file in your Streamlit app. Maintaining your API client outside your Streamlit app gives you reusability.

Copy the client file from the *chapter9/notebooks* directory (or the *chapter9/complete/ notebooks* directory if you haven't completed Chapter 9 yet) into the *chapter11/ streamlit* directory:

```
.../chapter11 (main) $ mkdir streamlit
.../chapter11 (main) $ cp ../chapter9/notebooks/swc_simple_client.py streamlit
```

Creating Your Streamlit App

Your Streamlit app will include an *entrypoint file*, which is the file that Streamlit loads first. In this app, you will use the entrypoint file to set the initial configuration and create the page navigation. Your app will also include individual page files that perform the work.

Since your API is running in the terminal window, you will need to open a second terminal window and execute the following command:

```
.../chapter11 (main) $ mkdir streamlit
.../chapter11 (main) $ cd streamlit
.../streamlit (main) $ touch streamlit_football_app.py
.../streamlit (main) $ touch page1.py
.../streamlit (main) $ touch page2.py
```

Updating the Entrypoint File

The entrypoint file contains a couple of items that are fundamental to a Streamlit app. The first is `st.session_state`, which is the mechanism to share information between pages in the application. The next is `st.navigation`, which is how you make a multi-page app with a shared navigation bar.

Add the following code to *streamlit_football_app.py*, and replace the statement [insert your API base URL] with the base API URL from the API running locally:

```
import streamlit as st
import logging
import pandas as pd

if 'base_url' not in st.session_state:
    st.session_state['base_url'] = "[insert your API base URL]"  ❶

logging.basicConfig(  ❷
    filename='football_app.log',
    level=logging.INFO,
)

st.set_page_config(page_title="Football App",
                   page_icon=":material/sports_football:")  ❸

page_1 = st.Page("page1.py", title="Team Rosters", icon=":material/trophy:")  ❹

page_2 = st.Page("page2.py", title="Team Stats", icon=":material/star_border:")

pg = st.navigation([page_1, page_2])  ❺
pg.run()  ❻
```

❶ Replace the statement [insert your API base URL] with the base API URL from the API running locally.

❷ This sets up the logging configuration for the application. All logging messages at INFO level or more severe will be saved in this file.

❸ This sets the configuration for the entire application, including the title and icon to display in the browser tab.

❹ This command defines the settings for the first subpage, along with the title and icon to display in the navigation bar.

❺ This creates the navigation bar and orders the subpages.

❻ This statement executes the current page.

Running Your Streamlit App

To run your app, execute the following command in the terminal:

```
.../streamlit (main) $ streamlit run streamlit_football_app.py
```

You will see the display shown in Figure 11-3.

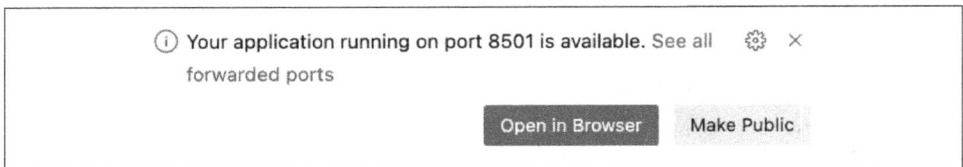

Figure 11-3. Launching your Streamlit app

Click "Open in Browser." You will see a blank browser window with a vertical ellipsis in the top right. Select the ellipsis and choose Settings. On the Settings screen select "Run on save," then close the Settings window to save.

If you are prompted with "Source code changed," select "Always rerun." This ensures that coding changes are applied to the running app immediately.

You will see the blank app with a navigation bar, as shown in Figure 11-4.

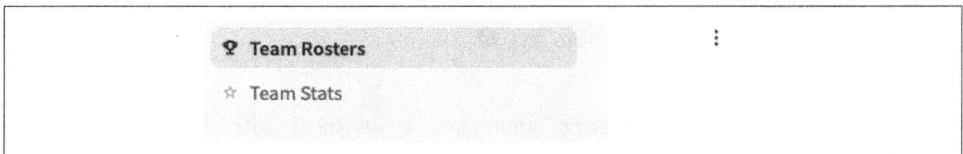

Figure 11-4. Initial blank data app

You can open the VS Code Simple Browser and have it running next to your code window. Open the VS Code Command Palette by clicking Ctrl-Shift-P. Enter **Simple Browser:Show,** then enter the address of the application (the one that ends with *app.github.dev*).

Creating the Team Rosters Page

The Team Rosters subpage will be displayed when your application executes. It will call the SportsWorldCentral API, manipulate the data using pandas, and then display the pandas DataFrame.

The first half of this file performs some library imports and sets up the logging. Then, it calls the API to retrieve all the data that will be used on the page for both filtering and display. The data is used to create a pandas DataFrame, which is a handy way of processing data in Python. The pandas library supports many powerful data manipulation functions, which you will see in the next section.

Add the following code to *page1.py*:

```
import streamlit as st
import swc_simple_client as swc
import pandas as pd
import logging

logger = logging.getLogger(__name__) ❶

st.header("SportsWorldCentral Data App") ❷
st.subheader("Team Rosters Page")

base_url = st.session_state['base_url'] ❸

try: ❹
    team_api_response = swc.call_api_endpoint(base_url,swc.LIST_TEAMS_ENDPOINT)❺

    if team_api_response.status_code == 200: ❻

        team_data = team_api_response.json() ❼

        teams_df = pd.DataFrame.from_dict(team_data) ❽

        unique_leagues = teams_df['league_id'].unique() ❾
        unique_leagues = sorted(unique_leagues.astype(str))

        if 'unique_leagues' not in st.session_state: ❿
            st.session_state['unique_leagues'] = unique_leagues
```

❶ This statement creates a reference to the logging file that was configured in the entrypoint file.

❷ These headings will be printed at the top of the page.

❸ This retrieves the `base_url` variable from the session state.

❹ All of the page's code is wrapped in a `try…except` structure. If any unhandled exception occurs, it logs the error and writes an error message to the screen.

❺ This uses *swc_simple_client.py* to call the API and stores the `httpx.response` in a variable.

❻ A `200` value in the `status_code` indicates a successful API call. It proceeds to populate the page.

❼ This converts the JSON data from the API to a Python representation of the data. The `.json()` name is a bit confusing since it doesn't return JSON data. It might help to think of this as the "converting from JSON" method.

❽ This uses pandas to create a DataFrame using the `from_dict` method. pandas DataFrames are convenient data structures for manipulating data.

❾ These two lines get a list of unique `league_id` values, convert them to a string, and sort them.

❿ This command stores the unique list of leagues in the `session_state` object so that all pages in the app can use them.

This page contains a navigation bar with a select box titled "Pick league ID." When users select a league that has rosters, the team rosters page will be filtered with teams that match that `league_id` value. This filtering is accomplished by filtering the Data-Frame and by Streamlit updating the display to match.

The remaining code creates the navigation bar, then uses pandas to format the data and display it on the page. It also contains matching error handling and exception handling statements from the previous code.

Add the following code to *page1.py*:

```
selected_league = st.sidebar.selectbox('Pick league ID',unique_leagues)❶

st.sidebar.divider()
st.sidebar.subheader(":blue[Data sources]")
st.sidebar.text("SportsWorldCentral")

flat_team_df = pd.json_normalize(
    team_data, 'players', ['team_id', 'team_name','league_id']) ❷
column_order = ['league_id','team_id','team_name','position',
                'player_id', 'gsis_id', 'first_name', 'last_name']
flat_team_df_ordered = flat_team_df[column_order]

if 'flat_team_df_ordered' not in st.session_state:
    st.session_state['flat_team_df_ordered'] = flat_team_df_ordered

display_df = flat_team_df_ordered.drop(columns=['team_id','player_id'])

display_df['league_id'] = display_df['league_id'].astype(str)
display_df = display_df[display_df['league_id'] == selected_league] ❸

st.dataframe(display_df,hide_index=True) ❹

    else:
        logger.error(f"Error encountered: {team_api_response.status_code}
                     {team_api_response.text}") ❺
        st.write("Error encountered while accessing data source.")

except Exception as e:
    logger.error(f"Exception encountered: {str(e)}") ❻
    st.write(f"An unexpected error occurred.")
```

❶ Create a select box in the navigation bar to select a league_id value.

❷ Use the pandas json_normalize() function to reformat the nested JSON data into rows and columns. Then create a new variable with a different column order.

❸ Create a filter DataFrame that contains only the values matching selected_league, which the user chose in the select box.

❹ Use Streamlit's built-in DataFrame() function to display the filtered DataFrame on the page.

❺ This will display an error if the API call is not successful and returns an HTTP status code other than 200.

❻ This is a general exception handling block that will be entered if anything goes wrong in the main processing of the page.

As you rerun the code on this page, you will see the display of the Team Rosters page, as shown in Figure 11-5.

	SportsWorldCentral Data App					
♟ Team Rosters	**Team Rosters Page**					
☆ Team Stats						
	league_id	team_name	position	gsis_id	first_name	last_name
Pick league ID	5001	Roaring Kitties	QB	00-0023459	Aaron	Rodgers
5001 ⌄	5001	Roaring Kitties	K	00-0023853	Matt	Prater
	5001	Roaring Kitties	TE	00-0024243	Marcedes	Lewis
Data sources	5001	Roaring Kitties	WR	00-0026293	Matt	Slater
SportsWorldCentral	5001	Roaring Kitties	WR	00-0027944	Julio	Jones
	5001	Roaring Kitties	RB	00-0028063	Taiwan	Jones
	5001	Roaring Kitties	RB	00-0029239	Brandon	Bolden

Figure 11-5. Team Rosters page

Creating the Team Stats Page

The Team Stats page will be displayed when users select Team Stats in the navigation bar. Instead of calling the SportsWorldCentral API, it reuses the data that was stored in `SessionState` as a pandas DataFrame.

The first portion of this file imports libraries and sets up the logging. Next, it creates items for the sidebar and filters the dataset to match the selected entry in the select box.

Add the following code to *page2.py*:

```
import streamlit as st
import pandas as pd
import logging
import nfl_data_py as nfl
import matplotlib.pyplot as plt ❶


logger = logging.getLogger(__name__)

st.header("SportsWorldCentral Data App")
st.subheader("Team Touchdown Totals")

try:
```

```
flat_team_df_ordered = st.session_state['flat_team_df_ordered'] ❷

unique_leagues = st.session_state['unique_leagues'] ❸
selected_league = st.sidebar.selectbox('Pick league ID',unique_leagues)

st.sidebar.divider()
st.sidebar.subheader(":blue[Data sources]")
st.sidebar.text("SportsWorldCentral")
st.sidebar.text("NFLDataPy")


flat_team_df_ordered['league_id'] = flat_team_df_ordered[
    'league_id'].astype(str) ❹
flat_team_df_ordered = flat_team_df_ordered[
    flat_team_df_ordered['league_id'] == selected_league]
```

❶ This library will be used to format a bar chart on this page.

❷ This retrieves the API data that was created in the Team Rosters page.

❸ This retrieves the unique_leagues that was created on the Team Rosters page and adds a select box to the sidebar.

❹ These filter the DataFrame to match the league_id from the select box in the navigation bar.

You are not limited to the data available in the SportsWorldCentral API— you can combine it with external sources. For the Team Stats page, you will enrich your data from the nfl_data_py library by joining on the GSIS_ID value. This allows your app to build a bar chart with touchdown data, which was not available in the API.

> In Part I of this book, you chose to include the gsis_id in your API, which comes from the NFL's Game Statistics and Information System. You didn't need it for your internal operations, but you decided it would benefit data science users who wanted to join your data to external sources. That extra thought is going to pay off now for your users, because it allows you to join your API data to data from nfl_data_py.

The remaining code in this page first loads the data from nfl_data_py into a pandas DataFrame, selects four columns, then creates a new total_tds column. Next, the code uses the merge command to join the API data to the nfl_data_py data, aggregates it, and displays a chart.

Add the following code to *page2.py*:

```
nfl_data_2023_df = nfl.import_seasonal_data([2023], 'REG') ❶

columns_to_select = [
    'player_id', 'passing_tds', 'rushing_tds', 'receiving_tds'] ❷
nfl_data_2023_subset_df = nfl_data_2023_df[columns_to_select].copy()

nfl_data_2023_subset_df['total_tds'] = ( ❸
    nfl_data_2023_subset_df['passing_tds'] +
    nfl_data_2023_subset_df['rushing_tds'] +
    nfl_data_2023_subset_df['receiving_tds']
)

merged_df = pd.merge(  ❹
    flat_team_df_ordered,
    nfl_data_2023_subset_df,
    how='left',
    left_on='gsis_id',
    right_on='player_id'
)

grouped_df = merged_df.groupby('team_name')['total_tds'].sum() ❺

fig, ax = plt.subplots() ❻
grouped_df.plot(kind="barh", xlabel='Total TDs',
                ylabel="Team Name", title="Total TDs - 2023", ax=ax)

st.pyplot(fig) ❼


except Exception as e:
    st.write(f"An unexpected error occurred: {str(e)}")
```

❶ This command calls the `import_seasonal_data` method from nfl_data_py and requests 2023 regular season data.

❷ The next two lines build a list with the column names you want and create a DataFrame with the columns identified in the previous step.

❸ These add a new calculated field to the DataFrame with the combined touchdowns for the team.

❹ This statement uses the `pandas.merge()` function, which allows you to join two DataFrames. You are merging the API DataFrame and the nfl_data_py DataFrame.

❺ The pandas `groupby` statement combines all the players for each team and sums the value in the `total_tds` column.

❻ The next two statements build a bar chart using pandas and the matplotlib library.

❼ This uses Streamlit's `pyplot()` function to display the plot created in the previous steps.

You will see the display of the Team Rosters page, as shown in Figure 11-6.

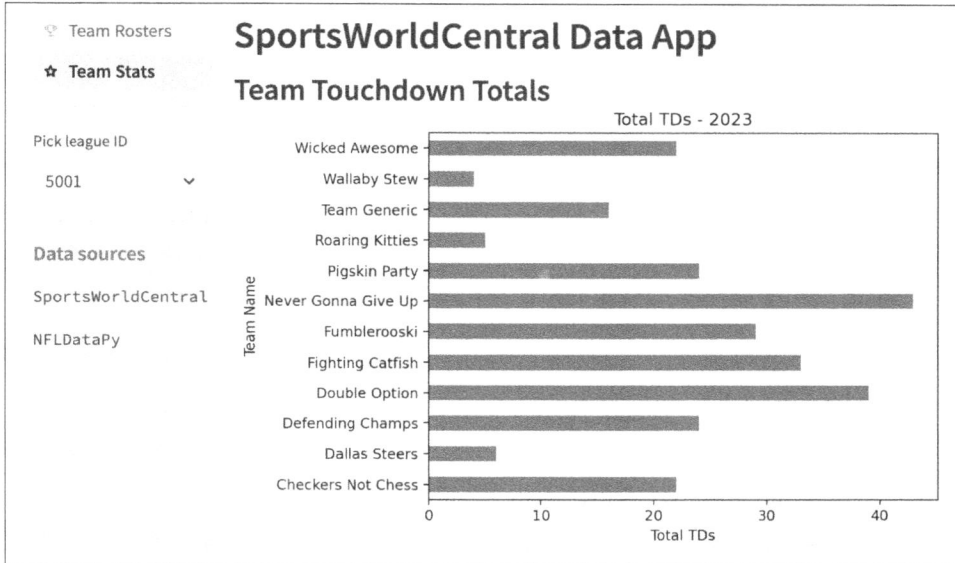

Figure 11-6. Team Stats page

To see the final structure of your Streamlit app, execute the `tree` command as follows:

```
.../streamlit (main) $  tree --prune -I 'build|*.egg-info|__pycache__'
.
├── football_app.log
├── page1.py
├── page2.py
├── streamlit_football_app.py
└── swc_simple_client.py

1 directory, 5 files
```

Congratulations, you've created a data app using Streamlit that calls the SportsWorld-Central API and combines it with NFL data from the nfl_data_py library.

Deploying Your Streamlit App

If you are going to show off your data science apps, you need to have them on the web. Streamlit Community Cloud is a free hosting option that allows you to deploy your app from your GitHub repository. One downside is that the app does not run continually—it periodically goes to sleep and the user has to wake it up. To deploy to Streamlit Community Cloud, follow the instructions from the Streamlit documentation (*https://oreil.ly/rJgTG*).

In addition to these two hosts, Streamlit links to other deployment tutorials (*https://oreil.ly/ZQXl4*) that you can follow. For the most part, these charge for hosting.

Completing Your Part II Portfolio Project

You have reached the end of Part II, congratulations! As you did with Part I, you will perform some housekeeping to get the portfolio project cleaned up and ready to share. You'll move the code out of the chapter folders into functional folders.

Before you make these changes, you'll save a copy of your files to a separate GitHub branch, named *learning-branch*, so that the files are still available if you want to continue working through the code.

Create the new branch from the command line as follows:

```
.../analytics-project/ (main) $ git checkout -b learning-branch ❶
Switched to a new branch 'learning-branch'
.../analytics-project/ (main) $ git push -u origin learning-branch ❷
 * [new branch]      learning-branch -> learning-branch
branch 'learning-branch' set up to track 'origin/learning-branch'.
```

❶ Create a new branch named *learning-branch* locally based on the *main* branch.

❷ Push this new branch to your remote repository on GitHub.com.

Next, you will make some changes to the directory structure. Enter these commands:

```
.../analytics-project/ (learning-branch) $ git checkout main ❶
Switched to branch 'main'
Your branch is up to date with 'origin/main'.
.../analytics-project/ (main) $ rm -rf chapter9/complete
.../analytics-project/ (main) $ rm -rf chapter10/complete
.../analytics-project/ (main) $ rm -rf chapter11/complete
.../analytics-project/ (main) $ mv chapter9/notebooks .
.../analytics-project/ (main) $ mkdir airflow ❷
.../analytics-project/ (main) $ mv chapter10/* airflow ❸
.../analytics-project/ (main) $ mv chapter11/streamlit . ❹
.../analytics-project/ (main) $ rm -rf chapter9 ❺
.../analytics-project/ (main) $ rm -rf chapter10
.../analytics-project/ (main) $ rm -rf chapter11
```

❶ Switch your Codespace back to the *main* branch of your repository.

❷ Make a new directory for the files from Chapter 10.

❸ Move Chapter 10's *airflow* files to the new folder.

❹ Move Chapter 11's *streamlit* folder to the root directory.

❺ Remove all the subdirectories and their files.

To see the directory structure of the completed project, run the following command:

```
.../portfolio-project (main) $ tree -d  --prune -I 'build|*.egg-info|__pycache__'
.
├── airflow
├── api
├── notebooks
└── streamlit

4 directories
```

Update the *README.md* file with the following at a minimum, then add your own thoughts about what you've learned in your project and how you've customized it:

```
# Analytics Portfolio Project
This repository contains programs using industry-standard Python frameworks,
based on projects from the book _Hands-on APIs for AI and Data Science_
written by Ryan Day.
```

Now commit these changes to GitHub, and your Part II portfolio project is ready to share with the world. Congratulations on completing your Part II capstone!

Extending Your Portfolio Project

Here are a few ideas to extend your Streamlit project:

- Implement Streamlit caching (*https://oreil.ly/_CVQ6*) in your API calls to improve performance in your application.
- Create additional visualizations with the built-in Streamlit chart elements (*https://oreil.ly/SdstJ*).

Additional Resources

If you'd like to learn more about the components of Streamlit, the Streamlit documentation (*https://docs.streamlit.io*) is well written and easy to use.

Tyler Richards' *Streamlit for Data Science, 2nd Edition* (Packt Publishing, 2023), is a thorough reference for using Streamlit in data science applications.

For a quick reference on Streamlit commands, view the official Streamlit Cheat Sheet (*https://oreil.ly/RirMd*).

Summary

In this chapter, you built an interactive data app with Streamlit. You consumed data from your SportsWorldCentral API and displayed that data in your app. Then, you combined it with NFL data from a Python library and created a chart with the data.

In Chapter 12, you will start learning how to use APIs with artificial intelligence.

Using APIs with Artificial Intelligence

In Part III, you will learn how APIs are used in artificial intelligence, including machine learning, the LangChain framework, and ChatGPT:

- Chapter 12 explores the relationship of AI and APIs and introduces your third portfolio project.
- In Chapter 13, you will create a machine learning model and deploy it as an API.
- Chapter 14 demonstrates calling an API using the LangChain and LangGraph frameworks.
- In Chapter 15, you will use OpenAI's ChatGPT to create a custom GPT and custom actions to interact with data from your API.

Using APIs with Artificial Intelligence

More AI means more APIs.

    —Frank Kilcommins, SmartBear

In technology circles, AI and APIs are sometimes treated as separate specialties. But they are closely related, and getting closer all the time. In this chapter, you will learn about the ways that AI and APIs overlap, some of the skills you should develop, and how to build APIs that are compatible with AI. Then, you will set up your Part III portfolio project, which you will use in the remaining chapters.

The Overlap of AI and APIs

To begin with, APIs are important data sources—along with databases and files—for training AI models. Once a model is trained, a REST API is a common method to make it available for users. You will train a machine learning model in Chapter 13, and deploy it with a REST API.

In the same way, cloud-based AI tools are advanced machine learning models that are deployed using APIs. AI tools such as generative AI, natural language processing, and others are often cloud hosted and made available as APIs. You will call an Anthropic large language model (LLM) through a REST API in Chapter 14.

An emerging area of overlap between AI and APIs is calling APIs directly from *generative AI* applications, which are built using LLMs and interact with users via natural language. One type of these applications is called retrieval augmented generation (RAG). In a RAG application, the program calls APIs and other data sources and then feeds the retrieved information to the LLM along with the user prompt. This helps overcome the knowledge gap, in which an LLM only has information that it was trained upon.

Another type of generative AI application uses LLMs to determine what API endpoints to use, which we will call *agentic* applications in this book. The LLMs make this decision by interpreting definitions from OAS files, Python code, or API documentation. You will create agentic AI applications that call APIs in Chapter 14 with LangChain and in Chapter 15 with ChatGPT.

API Perspectives: Bill Doerrfield on the Overlap of AI and APIs

Bill Doerrfield is the editor-in-chief at Nordic APIs (*https://nordicapis.com*), an international community of API practitioners and enthusiasts. Bill has been covering the API economy for more than a decade and therefore has a good perspective on how AI is impacting APIs.

How has AI affected the field of APIs in recent years?

I see the rise of AI and APIs as intrinsically tied. And this isn't just new to the era of ChatGPT and LLMs. We've seen APIs powering software-as-a-service products related to image recognition, object classification, natural language processing, speech-to-text, text-to-speech, chatbot frameworks, and more for years now. But more recently, as APIs have become the lingua franca for new AI services, we're seeing an interest in APIs as well.

What future trends do you see with the intersection of APIs and AI?

AI agents are progressively becoming more aware of APIs, and I see APIs as eventually connecting the dots behind the scenes of many complicated user-facing workflows. The next hurdle after that will be giving AI agents the power to not only integrate with data and make POST calls, but also access proprietary data and functionality that is typically monetized and perform transactions on behalf of the user.

Based on the rising importance of AI, what skills should API developers and designers be looking to develop?

API developers and designers should assume that anything that is publicly accessible will be tapped by an AI. RAG-based AI agents could also circumvent the need for APIs entirely, essentially like robotic process automation using screen scraping. So, we need to make APIs more accessible and defined for an LLM to research and learn about and for an agent to actually integrate with. Otherwise, developers will utilize these other means.

Developers should stay up-to-date with evolving trends in the API space, such as the OpenAPI Specification and its Arazzo Specification. They should consider using AI to enhance the API consumer's experience when possible. And API designers should also consider ensuring that their services are accessible where developers are already working, such as within their IDEs or AI assistants.

Designing APIs to Use with Generative AI and LLMs

So, how should you design an API so that a generative AI application can use it, as Doerrfield recommends? This field is changing rapidly, but here are some initial tips that apply to LangChain (Chapter 14) and ChatGPT (Chapter 15).

First, you need to consider whether an API or endpoint is appropriate to use with an agentic generative AI application without additional safeguards in place. The providers of LLMs provide warnings such as that LLMs should not be used "on their own in high-risk situations" (Anthropic Claude 3) and that they "may sometimes provide inaccurate information" (Google NotebookLM). ChatGPT's documentation simply admonishes the user to "check important info."

> Researchers and machine learning engineers are exploring additional methods to address risks of using LLMs for API calls and performing other business tasks. Some potential safeguards include requiring a human to approve tasks recommended by LLMs before executing, combining multiple AI agents to review tasks before executing, reviewing and filtering inputs and outputs to the models, and reviewing logs of the functioning of the system. In addition, foundational practices of API management and security are required when LLMs use APIs—just as they are when conventional software uses APIs. A key security practice is to restrict the permissions provided to systems that include LLMs.

For APIs that are appropriate, here are some design tips based on my own experience and on Blobr's "Is Your API AI-ready? Our Guidelines and Best Practices" (*https://oreil.ly/jxllA*):

Limit the size of the data results.

This is important for cost and accuracy. From a cost perspective, model providers charge for processing *tokens*, which are chunks of text. The size of these tokens differs, but the bottom line is the same: the more data a model processes, the greater the cost. In addition to the cost, developers using ChatGPT have found that it struggles to perform calculations from very large datasets returned by APIs. If you are using a model to perform calculations, limiting the size of the data results improves its accuracy.

There are a few ways to accomplish this. Rather than returning all fields related to an entity, return only the critical fields. If an API returns child records in a collection (e.g., `product.orders`), exclude these from the results and make them available in a separate endpoint. Add parameters, filters, and pagination to narrow down the specific records in the API call.

Make data structures consistent throughout the API.
 The more predictable the API is, the more accurately an AI can use it. By re-using schemas inside your APIs and defining them in your OAS file, you will help the LLM know what to expect in the results. You used Pydantic in the API you created in Part I, which enforced standard schemas and published them in your OAS file.

Provide a software development kit (SDK).
 Providing an SDK is a way to provide a subset of endpoints and customized API calls that are appropriate for an AI application. The SDK can also include detailed explanations of the API calls and parameters that assist an LLM in understanding its usage. In Chapter 14, you'll use the swcpy SDK with Lang-Chain and LangGraph.

Customize your OpenAPI Specification (OAS).
 Some methods of using generative AI support reading the OAS file to infer API endpoints to call. You can create a customized OAS file with AI-appropriate end-points and detailed descriptions of each endpoint and parameter that assist the LLM in inferring their meaning. Endpoints in the OAS file should have unique and clear operation IDs.

Provide a separate endpoint for summary statistics.
 If users ask the AI questions about summary information and counts, the LLM's behavior can be erratic. It may try to perform a scan of every record in the API, it may just look at the record identifiers and infer this is the count, or it may try something completely different. Providing dedicated endpoints takes some of the guesswork out.

Provide a search endpoint that doesn't rely on a record identifier.
 LLMs are more comfortable using language than numbers. They like to search based on the information that users are likely to ask them.

Arazzo to Define Multistep Processes

Because AI agents are nondeterministic, it can be difficult to ensure that they use APIs correctly, especially when multiple API calls are required to complete a business process. One current method of encouraging AI applications to use multiple API calls together is to add information to the descriptions of each API endpoint in the OAS file or tool functions. These descriptions can explain to AI models how one endpoint relates to another. But it is still up to the model to use them correctly.

An emerging trend that Doerrfield mentioned that may help with this challenge is the Arazzo Specification (*https://oreil.ly/qXKCU*). Arazzo defines "sequences of calls and their dependencies" so that multiple API endpoints can be used together to complete a business outcome. If a set of related calls is defined in Arazzo, it could be a deterministic building block that could be used by AI applications, adding reliability to API usage. Arazzo works with OpenAPI, so it is a natural fit for applications and tools that use OAS today.

Keep an eye on Arazzo to see how AI frameworks add formal support for it in the future. And it may be worth experimenting with LLMs today to see how they can use it today.

Defining Artificial Intelligence

Artificial Intelligence is technology that enables computers and machines to simulate human learning, comprehension, problem solving, decision making, creativity and autonomy.

—"What Is Artificial Intelligence (AI)?", Cole Stryker and Eda Kavlakoglu, IBM Corporation, 2024

Aside from that formal definition of AI, an informal definition today is that AI is a computer program that can have humanlike conversations and complete humanlike tasks. AI includes *expert systems*, which have been around for decades. These are complicated rules-based systems that can perform humanlike tasks. Modern AI focuses on *machine learning*, in which researchers direct the training of models but don't explicitly program them. Generative AI using LLMs is one major application of machine learning.

Figure 12-1 demonstrates how these terms relate to one another.

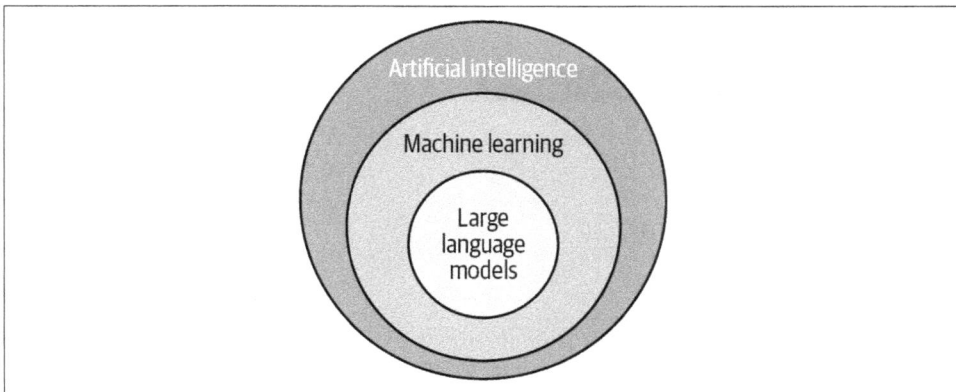

Figure 12-1. Diagram of AI terminology

Generative AI and Large Language Models (LLMs)

Although machine learning is used in many different applications, most of the attention from the general public in recent years has been given to generative AI. The ability of these applications to generate text, music, and videos based on text prompts has led to rapid adoption in applications such as OpenAI's ChatGPT, Microsoft's Copilot, Google's Gemini, and many additions to other software applications.

Despite the impressive capabilities of these applications, generative AI also has many risks and limitations associated with it. Providers of some popular models include warnings about bias, hallucinations, mistakes, and harmful content. These are major risks that should be taken seriously by developers who use them. Chapters 14 and 15 have more details on the risks and limitations of the models demonstrated in those chapters.

Creating Agentic AI Applications

As Doerrfield mentions, AI agents are at the forefront of AI research and development. An *agent* is software that controls application flow using an LLM. The more autonomously the LLM controls the system, the more *agentic* the system is.

Creating AI agents using LLMs is a new field, and a variety of different tools have been released to create agents or orchestrate multiple agents to perform tasks. Table 12-1 lists several open source frameworks for developing agents and LLM-based applications.

Table 12-1. AI agent frameworks

Software	Programming languages supported
Autogen	Python, dotnet
CrewAI	Python
LangChain/LangGraph	Python
LlamaIndex	Python, Typescript
PydanticAI	Python
Vercel AI SDK	Typescript

You will use LangChain and LangGraph in Chapter 14 to call APIs.

AI Perspectives: Samuel Colvin on Agentic AI

Samuel Colvin is the founder of Pydantic, the open source Python library that you have been using throughout this book for data validation and serialization. Pydantic is used in many of the model providers' SDKs and Python-based AI agent frameworks. I talked to Colvin about agents and Pydantic's own agent framework, PydanticAI.

What is the importance of AI agents and agent frameworks in the next few years of AI development?

I'm confident that in time, people won't be using the AI providers' SDKs directly. There will be some libraries on top that make opinionated decisions about what people want to do. If you look at web frameworks as a parallel, technically they are constraints on what you can do [to create web applications], but they are constraints that everyone is happy with because you get to write much more high-level code. I'm hopeful that PydanticAI is a big part of that.

What role do you see PydanticAI playing in creating AI applications?

My hope is that PydanticAI can become the default in terms of how you actually take GenAI and put it into production, because we have had more Python experience but also just work harder on the high-level developer experience and the kinds of checks you need when you're trying to put that stuff into production. For example, for PydanticAI we have put enormous effort into making it type safe, including the dependency system.

With the AI era that we're moving into, what are the skills or tools or frameworks you think developers should be learning right now?

I think that there are bits of intuition about how LLMs work, nuances of the peculiarities of how they behave, that are worth getting to grips with. And good fundamental Python programming techniques: there is a temptation to think I can skip learning the fundamentals of good programming techniques because I'm doing GenAI. If anything, some of those things are even more important. The fact that you have a nondeterministic system that you're interacting with means you need to be even more sure that your fundamental unit tests for the bits that should be deterministic are rock solid.

Two more standards that Colvin recommends developers becoming familiar with are OpenTelemetry (*https://oreil.ly/OTEL*) for observability and logging of AI applications, and Model Context Protocol (MCP) (*https://oreil.ly/mcptL*) for providing context to LLMs.

Introducing Your Part III Portfolio Project

You will create a portfolio project that demonstrates your ability to work with API and AI. Here is an overview of the work ahead of you:

- Chapter 13: Deploying a machine learning API
- Chapter 14: Using APIs with LangChain
- Chapter 15: Using ChatGPT to call your API

Each of these tasks will enable you to showcase your API and AI skills in a unique way.

Getting Started with Your GitHub Codespace

You will continue to use GitHub Codespaces for all the code you develop in Part III. If you didn't create a GitHub account yet, do that now.

Cloning the Part III Repository

All of the Part III code examples are contained in this book's GitHub repository (*https://github.com/handsonapibook/api-book-part-three*).

To clone the repository, log in to GitHub and go the GitHub Import Repository page (*https://github.com/new/import*). Enter the following information in the fields on this page:

- The URL for your source repository: **https://github.com/handsonapibook/api-book-part-three**
- Your username for your source code repository: Leave blank.
- Your access token or password for your source code repository: Leave blank.
- Repository name: **ai-project**
- Public: Select this so that you can share the results of the work you are doing.

Click Begin Import. The import process will begin, and the message "Preparing your new repository" will be displayed. After several minutes, you will receive an email notifying you that your import has finished. Follow the link to your new cloned repository.

Launching Your GitHub Codespace

In your new repository, click the Code button and select the Codespaces tab. Click "Create codespace on main." You should see a page with the status "Setting up your codespace." Your Codespace window will be opened as the setup continues. When the setup completes, your display will look similar to Figure 12-2.

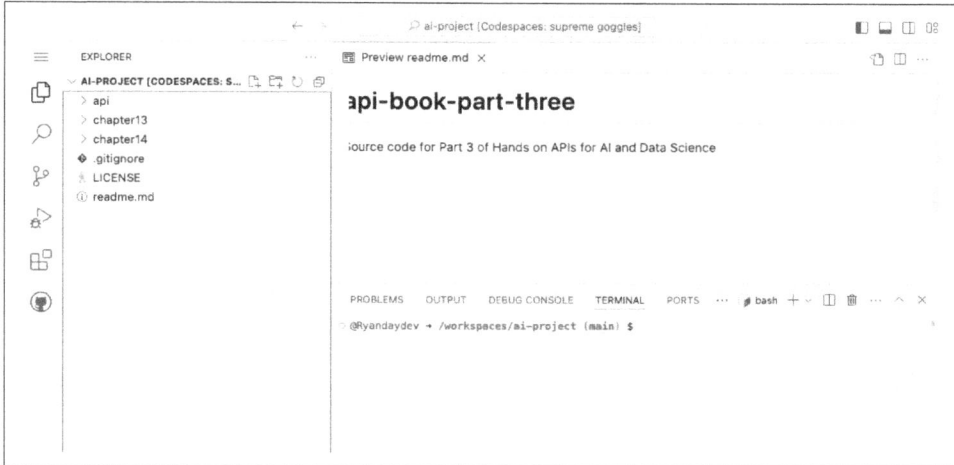

Figure 12-2. GitHub Codespace for Part III

Your Codespace is now created with the cloned repository. This is the environment you will be using for Part III of this book. Open the GitHub Codespaces page (*https://oreil.ly/nLbqH*) and scroll down the page to find this new Codespace, click the ellipsis to the right of the name, and select Rename. Enter the name **Part 3 Portfolio project codespace** and click Save. You should see the message "Your codespace *Part 3 Portfolio project codespace* has been updated." Click the ellipsis again and then click the ribbon next to "Auto-delete codespace" to turn off auto-deletion.

> To save space on the page, I have trimmed the directory listing in the terminal prompt of my Codespace. You can do this in your Codespace by editing the */home/codespace/.bashrc* file in VS Code. Find the export PROMPT_DIRTRIM statement and set it to export PROMPT_DIRTRIM=1. To load the values the first time, execute this terminal command: source ~/.bashrc.

Additional Resources

To view a 100-point scorecard for AI compatibility of APIs, read Blobr's "Is Your API AI-ready? Our Guidelines and Best Practices" (*https://oreil.ly/JLaK1*).

To see an example of working around the limitations of a GPT, read "Syntax Sunday: Custom API Wrapper for GPTs" by Kade Halabuza (*https://oreil.ly/khh0J*).

To learn more about possible futures of AI and APIs, read "AI + APIs — What 12 Experts Think The Future Holds" by Peter Schroeder (*https://oreil.ly/t2lxH*).

Summary

This chapter introduced the basics of AI and explained how it relates to APIs.

In Chapter 13, you will create a machine learning model and deploy it using FastAPI.

Deploying a Machine Learning API

Always in motion is the future.
　　—Yoda, *The Empire Strikes Back*

Fantasy football managers spend most of their time attempting to predict the future and plotting strategies based on those predictions. Before the season begins, managers want to know how NFL players will perform in the upcoming season so that they can build the best team. During their fantasy drafts, managers want to know where a player would be picked by other managers so that they can outmaneuver their competition. Each week, managers want to know which of their players are going to score the most so that they can set their lineups accordingly.

Many fantasy websites and platforms provide predictions to these managers. One of the tools available to the platforms is a *machine learning* (ML) model, which you learned about in Chapter 12. The platforms train various models and use them to make predictions, or *inferences*, to managers. If a model processes an entire group of predictions at once, it is called *batch inference*. Some fantasy questions are appropriate for batch inference, such as making a week's worth of player predictions all at once. Batch inference may be done by a scheduled script or job. But if the predictions are changing minute by minute—like in the case of a live score prediction for a game— then real-time inference is needed. *Real-time inference* is calling a model to get a single prediction immediately. This is where deploying the model as an API is most valuable.

In this chapter, you will create an ML model and deploy it with an API to make real-time inference. As you proceed through the chapter, here are a few terms that you will come across:

Classification
> A type of model that predicts what category a value will fall into. For example, a classification model might predict if a player will be drafted or undrafted. Models that perform classification are called *classifiers*.

Decision trees
> A type of ML algorithm that creates a recursive tree structure to perform classification or regression.

Evaluating a model
> Comparing the model's predictions to test data to see how well it would have predicted past events.

Gradient boosting
> An ML technique that combines multiple models to create a model that is more effective than the individual models.

Regression
> A type of model that predicts a continuous numeric value. For example, a regression model might predict how many points a player will score. Models that perform regression are called *regressors*.

Training a model
> Using the training portion of historical data to create a model that can make inferences based on new data.

Training Machine Learning Models

Supervised learning is a method of creating models by processing existing data where the expected values are known. For example, a financial fraud detection model might be trained by processing a large number of bank transactions that have been *labeled* or categorized as either fraud or nonfraud. Through this process, the model recognizes future records that are potentially fraudulent. Through this type of supervised learning, ML models can be created that create predictions on various data formats including tabular data, images, audio files, and others.

Figure 13-1 shows this type of training.

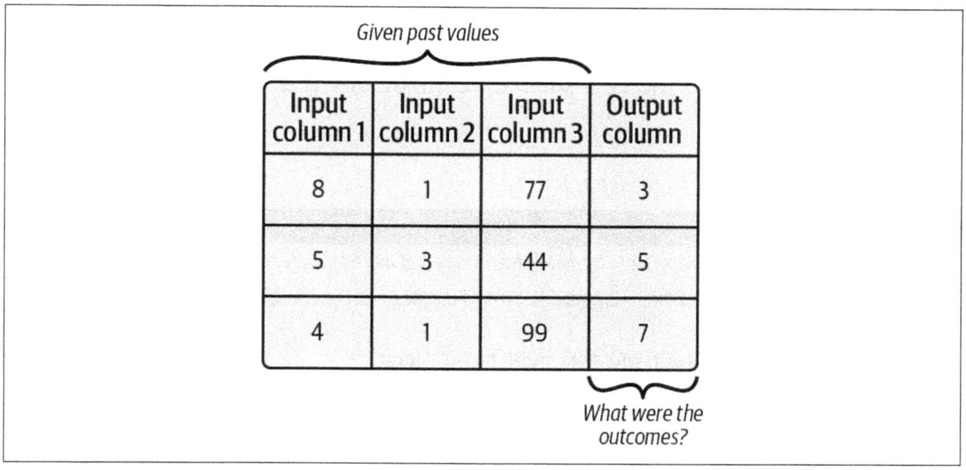

Figure 13-1. ML training model

The diagram shows a set of historical rows of data. The goal of the ML model in this case would be to predict future values of the output column. The training process would involve using software to read the input columns from historical data and look for patterns in how they are related to the output column.

When the model has been trained, it can be used to read the input columns from new rows of data and predict what the values will be for the output columns. This is the *inference* process, shown in Figure 13-2.

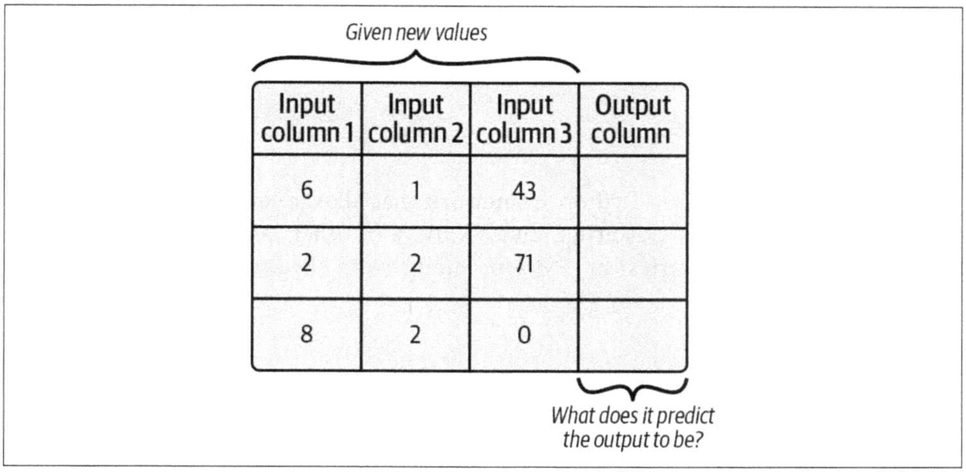

Figure 13-2. ML inference model

The model you will create in this chapter is a supervised ML model.

New Software Used in This Chapter

Table 13-1 lists a few of the new software components you will begin using in this chapter.

Table 13-1. Software used in this chapter

Software name	Purpose
ONNX Runtime	A cross-platform tool for using models from a variety of different frameworks.
scikit-learn	An ML framework for training models. You will use the `GradientBoostingRegressor` from this library.
sklearn-onnx	A library that converts scikit-learn models to ONNX format.

ONNX Runtime

The Open Neural Network Exchange (ONNX) is an open standard for ML models. Because such a variety of programming languages and libraries are used to make ML models, it can be complicated to deploy and run multiple different models. ONNX is a standard format that models from different programming languages and different frameworks can be converted to and run in a standard way.

This allows greater interoperability, because when models from different programming languages and frameworks are converted to ONNX format, they can be more easily deployed using the standard ONNX Runtime. The ONNX Runtime also includes acceleration that can improve model inference performance.

After you have developed your model in scikit-learn, you will convert it to ONNX format, and then use the ONNX Runtime (*https://oreil.ly/IGEBD*) in your API to make predictions (inferences).

scikit-learn

The scikit-learn library is a Python framework that allows you to create models for classification, regression, clustering, and a variety of other tasks. This is one of the more popular ML libraries in Python, along with PyTorch, TensorFlow, and XGBoost.

sklearn-onnx

Since you are using scikit-learn to create your model, you will use the sklearn-onnx library to convert your model into ONNX format. This will be the final step of the model training process.

Installing the New Libraries in Your Codespace

Open the Part III GitHub Codespace that you created in Chapter 12. To install the libraries you need for this chapter, create a file named *chapter13/requirements.txt*:

```
.../ai-project (main) $ cd chapter13
.../chapter13 (main) $ touch requirements.txt
```

Update *chapter13/requirements.txt* with the following contents:

```
#model training
scikit-learn ❶
numpy ❷
pandas ❸
skl2onnx ❹
pydantic>=2.4.0
fastapi[standard]>=0.115.0
uvicorn>=0.23.0 ❺
onnxruntime ❻
```

❶   The scikit-learn library will be used to create the ML model.

❷   The numpy library will be used to format numbers.

❸   The pandas library will be used to process the input data file.

❹   The skl2onnx library will be used to save the scikit-learn model into ONNX format.

❺   Uvicorn is the web server used to host FastAPI.

❻   The onnxruntime library is used to perform inference with a saved ONNX model file.

Execute the following command to install the new libraries in your Codespace:

```
.../chapter13 (main) $ pip3 install -r requirements.txt
```

You should see a message that states that these libraries were successfully installed.

Using the CRISP-DM Process

ML projects have many steps requiring people with a lot of specialized skills. A useful method of organizing an ML modeling project is the Cross-Industry Standard Process for Data Mining (Shearer, 2000). This model is widely used in the data science community.

The following are definitions of the stages in CRISP-DM:

Business understanding
> During the this stage, the team identifies business objectives and assesses tools and techniques available.

Data understanding
> Collecting data that is available to solve the problem, explore it, and verify the data quality.

Data preparation
> During this stage, data scientists select specific data elements to be used, format them, and merge with any additional sources needed.

Modeling
> Selecting a modeling technique and building a model that answers your business question.

Evaluation
> Review the model for its ability to solve the question and its readiness for production.

Deployment
> Models are deployed in an environment where they can be consumed by the customer. Monitor and maintain the model.

You will follow this process as you proceed with the chapter. The primary focus is on the deployment stage, so I will only touch lightly on some of the other stages.

Business Understanding

The first stage of the process is to establish a business understanding of the problem you are trying to solve. You are creating a model to serve fantasy football managers who are running their own team in a league with other owners. The question they need to answer each week of the season is "How much will it cost to acquire this player on waivers?"

Fantasy managers can add new players to their rosters through a *waiver request*. In many leagues, a blind bidding auction is performed to decide who gets the best available players. Managers decide which players they want to bid for and put in the dollar amount they want to spend, which is hidden from other managers. When the bidding is processed on Tuesday or Wednesday of each week, the highest bidder gets the player at full price. Lower bidders miss out (but also don't lose their money).

Each manager has a set amount of money they can use for the season, such as $100. (These aren't real-world dollars, these are fantasy dollars.) This is sometimes called the *free agent acquisition budget* (FAAB). A manager wants to bid high enough to win the bid, but not overspend. The best-case scenario would be to win the bid at a lower dollar amount—getting a bargain.

To help the manager bid enough to win the player they want without overspending, you will give the manager a range of predictions: the low-end cost (10th percentile), the median cost (50th percentile), and the high-end cost (90th percentile).

Data Understanding

In this stage, you will collect and explore the data that is available for your project. In the project repository, you'll find the file *player_training_data_full.csv*. It contains historical fantasy football transaction data with the following columns:

Fantasy regular season weeks remaining
How many weeks are left in the regular season. For example, in week 2 of a season with 14 weeks, this would be 12.

League budget percentage remaining
The percent of total dollars available in the league. For example, if $900 remain in the league's original $1,200, this would be 75.

Player season number
The number of seasons this player has been in the league. Rookies have a value of 1.

Position
The fantasy football position of the players that was acquired.

Waiver value tier
A qualitative measure of how valuable an individual player is. Each week, some players are "top tier" pickup targets, and they would get a 1. Players who are nothing special would get a 5. This is a categorical feature because putting players into the tiers is a qualitative judgment. (You may get these from a fantasy website or assign them yourself.)

To begin reviewing the data and selecting fields you want to use in your model, create a Jupyter Notebook by running the following commands in the Terminal window:

```
.../ai-project (main) $ cd chapter13
.../chapter13 (main) $ touch player_acquisition_model.ipynb
```

Open the *player_acquisition_model.ipynb* file. As you did in Chapter 9, select the Python kernel and enable the Python and Jupyter extensions, then select the recommended Python environment.

Enter the following title in the Markdown cell and run it:

```
# Player Acquisition Models
*This notebook is used to train a machine learning model using scikit-learn
and save it in ONNX format.*
```

Now you will import the Python libraries you need. Create and run the following Markdown cell:

```
## Library imports
```

Add and run a new Python cell with the following code:

```
import logging
import numpy as np
import pandas as pd
from sklearn.model_selection import train_test_split ❶
from sklearn.ensemble import GradientBoostingRegressor ❷

from skl2onnx import to_onnx
import onnxruntime as rt
```

❶ This function is used to split data files into train and test sets.

❷ Your model will use the GradientBoostingRegressor algorithm.

Add another Markdown cell with the following text:

```
## Configure logging
```

Add and run a Python code cell with the following:

```
for handler in logging.root.handlers[:]: ❶
    logging.root.removeHandler(handler)

logging.basicConfig(
    filename='player_acquisition_notebook.log',
    level=logging.INFO, ❷
)
```

❶ This statement removes any existing logging handlers configured by CodeSpaces.

❷ This sets the logging level to record in the log.

To begin loading your training data, add another Markdown cell with the following text:

```
## Load data
```

Add and run a Python code cell with the following:

```
dataset=pd.read_csv("player_training_data_full.csv")
```

Data Preparation

Next you will select the data to be included in the model. Rather than simply trying out all possible variables, you should consider the reason or theory that each would make a contribution to your model. Here are the three columns you'll include, and the theory behind each:

League budget percentage remaining
> Your intuition is that a higher budget remaining leads to higher bids. This would make this a *linear* feature, in which the output variable goes up or down at a consistent rate as this value changes.

Fantasy regular season weeks remaining
> The theory here is that players cost more at different points of the season. This probably isn't a strictly linear value. History suggests some of the highest bids come in at the beginning of the season when the starting lineups are revealed, but other peak bids occur from injured players during the season and when managers have "use it or lose it" at the end of the season.

Waiver value tier
> At a high level this is straightforward: higher-value players will cost more. But how much more? And how is each tier affected? These are more nuanced questions that you hope the model will be able to detect in the training data.

Modeling

Now you will begin the modeling stage, first by selecting the algorithm and ML framework to use for your model. This decision is a combination of technical limitations and modeling factors.

Your technical limitations are that you want to use a Python framework and you want to convert the model to the ONNX format for inference. You also want to make predictions for the 10th, 50th, and 90th percentiles, so you need to use an algorithm that meets these technical criteria.

Two modeling factors to consider are the type of output and the features you've selected. Your output will be numerical dollar values, so you will use a regression model (regressor). If your input features were all linear (as your input goes up or down, your prediction goes up or down), you could use a linear regressor. But your selected features are budget remaining (a linear feature), value tier (a categorical feature), and weeks remaining (a slightly more complicated feature). Because of the complexity of these features, some type of decision tree regressor is more appropriate.

Based on these technical and modeling factors, you will use the `GradientBoostingRe gressor` algorithm (*https://oreil.ly/_gqSO*) from scikit-learn. The gradient boosting algorithm is way of combining multiple decision trees into an ensemble model that is more predictive than using the individual decision trees by themselves. It also supports the multiple predictions by percentile that you want to use. This algorithm is also supported by the ONNX format you will be saving the model in.

To get started with the modeling process, you will first split your data into multiple variables for the training (80% of the rows) and testing (20% of the rows). There are conventions for naming of variables, and you will follow those so that your code is understandable by other data scientists. Table 13-2 explains the purpose of these variable names.

Table 13-2. Conventional variable names for training models

Variable name	Purpose	Columns included	Data included
X (uppercase)	Input columns for full data	Input	All
y (lowercase)	Output columns for full data	Output	All
X_train	Input columns of training data	Input	Training data (80%)
y_train	Output columns of training data	Output	Training data (80%)
X_test	Input columns of test data	Input	Test data (20%)
y_test	Output columns of test data	Output	Test data (20%)

Add another Markdown cell with the following text:

```
## Prepare and split data
```

Add and run a Python code cell with the following:

```
X = dataset[['waiver_value_tier','fantasy_regular_season_weeks_remaining',
            'league_budget_pct_remaining']] ❶
y = dataset['winning_bid_dollars'] ❷
X_train, X_test, y_train, y_test = train_test_split(X, ❸
                                                    y,
                                                    test_size = 0.20, ❹
                                                    random_state = 0) ❺
```

❶ You are selecting three of the input columns for X.

❷ You only include the output column when creating y.

❸ The `train_test_split` function reads the X and y variables and then outputs the variables explained in Table 13-2.

❹ This parameter determines that 20% of the data will be in the test set and 80% will be in the training set.

❺ If you use the same `random_state` value each time you call this method, you will get the same rows in the train and test variables.

Now that you split the data, you are ready to build a model. Because you want to give a range of predictions, you will create three separate models. When you create the API, you will combine the results of these models into a single API call.

The process of training your model is called *fitting*, where the library takes a general algorithm and *fits* it or applies it to your training data to make a specialized model.

Add another Markdown cell with the following text:

```
## Creating and fitting models
```

Add and run a Python code cell with the following:

```
model_10th_percentile = GradientBoostingRegressor(loss="quantile", alpha=0.1) ❶
model_50th_percentile = GradientBoostingRegressor(loss="quantile", alpha=0.5)
model_90th_percentile = GradientBoostingRegressor(loss="quantile", alpha=0.9)

model_10th_percentile.fit(X_train, y_train) ❷
model_50th_percentile.fit(X_train, y_train)
model_90th_percentile.fit(X_train, y_train)
```

❶ This command creates a `GradientBoostingRegressor` model that will try to predict the 10th percentile values. In this case, this means a dollar amount that will be less than 90% of the bids. The next two statements are similar except with different percentiles.

❷ This statement uses the `fit()` method to prepare this model to make predictions based on the training data you provided. The next two lines do the same for the other two models.

At this point, your models are in scikit_learn format and are only available in this Jupyter Notebook. To prepare these models for deployment and make them more cross-platform compatible, you will save your models in the ONNX format. Before doing this, you'll need to combine the features from the X variable into the two-dimensional array format required by the converter. Add another Markdown cell with the following text:

```
## Convert and save these models in ONNX format
```

Add and run a Python code cell with the following:

```
X_array = np.column_stack((X['waiver_value_tier'], ❶
                           X['fantasy_regular_season_weeks_remaining'],
                           X['league_budget_pct_remaining']))

onx = to_onnx(model_10th_percentile, X_array[:1]) ❷
with open("acquisition_model_10.onnx", "wb") as f:❸
```

```
    f.write(onx.SerializeToString())

onx = to_onnx(model_50th_percentile, X_array[:1])
with open("acquisition_model_50.onnx", "wb") as f:
    f.write(onx.SerializeToString())

onx = to_onnx(model_90th_percentile, X_array[:1])
with open("acquisition_model_90.onnx", "wb") as f:
    f.write(onx.SerializeToString())
```

❶ This statement combines the features from the X variable into the two-dimensional array format required by the convertor.

❷ This statement converts the first model to ONNX format. It sets the names of the input and output attributes in the model by reading the first row of the X_array, which contains the element names.

❸ This statement creates a file in the local filesystem and saves the model in ONNX format.

Evaluation

Planning and training models is an iterative process, so the model at this point likely needs improving. In a full project, you would iteratively evaluate the model with formal metrics for accuracy, fairness, and other qualities that make a model appropriate for production. At that point, you might decide to try a different combination of features and tune your model in different ways.

Since this chapter is focused on deploying models, you will not be performing those steps. For more information about model evaluation, read *Designing Machine Learning Sytems* by Chip Huyen (O'Reilly, 2022).

Deployment

You are are ready to deploy the model for real-time inference, with one API call returning a prediction that combines all three models.

Figure 13-3 demonstrates the components used to create this API. If you compare this to the components of the API created in Part I of this book, you will see many similarities. FastAPI is still used as the controller, and Pydantic is still used for data transfer and data validation. However, instead of retrieving data from a database like the Part I API did, this API will use the ONNX Runtime to perform inference from the models that you trained and saved.

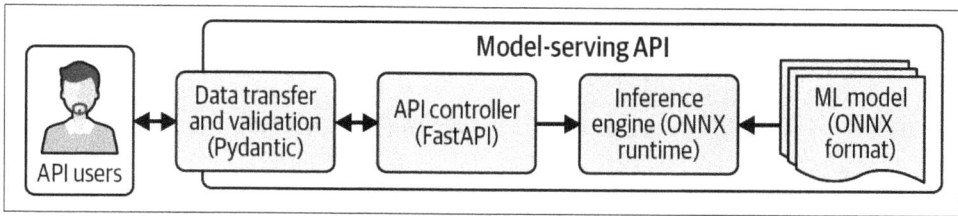

Figure 13-3. Model-serving API components

Begin by creating a Pydantic file named *schemas.py* to define the inputs and outputs to the API. FastAPI will use these schemas to generate the OAS file:

```
.../chapter13 (main) $ touch schemas.py
```

Add the following to this file:

```
"""Pydantic schemas"""
from pydantic import BaseModel ❶

class FantasyAcquisitionFeatures(BaseModel): ❷
    waiver_value_tier: int
    fantasy_regular_season_weeks_remaining: int
    league_budget_pct_remaining: int

class PredictionOutput(BaseModel): ❸
    winning_bid_10th_percentile: float
    winning_bid_50th_percentile: float
    winning_bid_90th_percentile: float
```

❶ The Pydantic library includes a `BaseModel` object that contains the validation logic.

❷ This class defines the input values that users will send to get a prediction from the model.

❸ This class defines the output that will be returned from the model. It contains three values—one from each model that you trained in the previous section.

Next, create the *main.py* file, which will contain the rest of the code for your API:

```
.../chapter13 (main) $ touch main.py
```

At the top of this file, add the imports and an API description. These will be used in the OAS file and then displayed on the Swagger UI documentation that FastAPI produces. Add this Python code:

```
"""Fantasy acquisition API"""

from fastapi import FastAPI
import onnxruntime as rt ❶
```

```
import numpy as np
from schemas import FantasyAcquisitionFeatures, PredictionOutput  ❷

api_description = """
This API predicts the range of costs to acquire a player in fantasy football

The endpoints are grouped into the following categories:

## Analytics
Get information about health of the API.

## Prediction
Get predictions of player acquisition cost.
"""
```

❶   This library is used to load the models from their files and serve inferences in the API.

❷   This imports the Pydantic schemas, which will be used to define the inputs and outputs of the API.

Next, you will add the code that uses the ONNX Runtime to load an inference session object for each of the three models. Then, these sessions are used to get labels for the input and output expected for this model. You defined the three inputs expected and the one output when you created the model in scikit-learn and then converted it to ONNX format.

Add the following code to the bottom of *main.py*:

```
# Load the ONNX model
sess_10 = rt.InferenceSession("acquisition_model_10.onnx",
                              providers=["CPUExecutionProvider"])  ❶
sess_50 = rt.InferenceSession("acquisition_model_50.onnx",
                              providers=["CPUExecutionProvider"])
sess_90 = rt.InferenceSession("acquisition_model_90.onnx",
                              providers=["CPUExecutionProvider"])

# Get the input and output names of the model
input_name_10 = sess_10.get_inputs()[0].name   ❷
label_name_10 = sess_10.get_outputs()[0].name  ❸
input_name_50 = sess_50.get_inputs()[0].name
label_name_50 = sess_50.get_outputs()[0].name
input_name_90 = sess_90.get_inputs()[0].name
label_name_90 = sess_90.get_outputs()[0].name
```

❶   This loads the first ONNX model from the file and creates a session object that can be used to make inferences. The next two lines do the same for the other model files.

❷ This statement gets the name of the input features from the session object. These will be used when making inferences.

❸ This statement gets the name of the output features from the session object. These will be used when making inferences.

Because you have placed this code outside any function definitions, it will run once at startup of the API.

The next section of FastAPI code will be familiar to you if you created the API in Part I. The first statement creates the FastAPI `app` object using the API description you added previously. Then, the `@app.get()` method creates the health check. This is a useful best practice that allows users to check the status of the API before making other API calls.

Add the following code to the bottom of *main.py*:

```
app = FastAPI( ❶
    description=api_description,
    title="Fantasy acquisition API",
    version="0.1",
)

@app.get( ❷
    "/",
    summary="Check to see if the Fantasy acquisition API is running",
    description="""Use this endpoint to check if the API is running. You can
    also check it first before making other calls to be sure it's running.""",
    response_description="A JSON record with a message in it. If the API is
    running the message will say successful.",
    operation_id="v0_health_check",
    tags=["analytics"],
)
def root(): ❸
    return {"message": "API health check successful"}
```

❶ This creates the main FastAPI `app` object using the `api_description` defined previously.

❷ This is the FastAPI decorator that defines a GET endpoint at the root address.

❸ This is the function that will be executed when this endpoint is called. It returns a single Python statement to show that the API is running.

The remaining code defines the API endpoint that provides the prediction capabilities for users. It begins with a Python decorator that provides information that will end up in the OAS file (and documentation). Then, it has the `predict()` method that uses the ONNX Runtime to call each model and put their outputs in the API response.

Add the following code to the bottom of *main.py*:

```
# Define the prediction route
@app.post("/predict/",  ❶
          response_model=PredictionOutput,❷
          summary="Predict the cost of acquiring a player",
          description="""Use this endpoint to predict the range of cost to
          acquire a player in fantasy football.""",
          response_description="""A JSON record three predicted amounts.
          Together they give a possible range of acquisition costs for a
          player.""",
          operation_id="v0_predict",
          tags=["prediction"],
)
def predict(features: FantasyAcquisitionFeatures):  ❸
    # Convert Pydantic model to NumPy array
    input_data = np.array([[features.waiver_value_tier,  ❹
                            features.fantasy_regular_season_weeks_remaining,
                            features.league_budget_pct_remaining]],
                            dtype=np.int64)

    pred_onx_10 = sess_10.run([label_name_10], {input_name_10: input_data})[0]  ❺
    pred_onx_50 = sess_50.run([label_name_50], {input_name_50: input_data})[0]
    pred_onx_90 = sess_90.run([label_name_90], {input_name_90: input_data})[0]


    # Return prediction as a Pydantic response model
    return PredictionOutput(winning_bid_10th_percentile=round(
                            float(pred_onx_10[0]),2),  ❻
                            winning_bid_50th_percentile=round(
                            float(pred_onx_50[0]),2),
                            winning_bid_90th_percentile=round(
                            float(pred_onx_90[0]), 2))
```

❶ This decorator creates a POST endpoint at the */predict* address. It will be used to perform inferences.

❷ The ResponseModel statement is used by FastAPI to define the return type of this endpoint. This will be used to generate the OAS file.

❸ This is the function that will be called at this endpoint.

❹ This statement reformats the input variables into a NumPy array, which is expected by the ONNX Runtime to call the model.

❺ This statement calls the ONNX Runtime and gets an inference for the 10th percentile model using the input from the API call. The next two statements use the same input to call the other two models.

❻ This statement creates a `PredictionOutput` object with the inference values, and returns it in the API call. It rounds the values to two decimal places for presentation.

With all of the API code completed, you are ready to run the API and test out the ML model. Enter the following command from the command line:

```
.../chapter13 (main) $ fastapi run main.py
```

You will see the application startup occur as shown in Figure 13-4.

```
module    🗲 main.py

code      Importing the FastAPI app object from the module with the following code:

          from main import app

app       Using import string: main:app

server    Server started at http://0.0.0.0:8000
server    Documentation at http://0.0.0.0:8000/docs

          Logs:

INFO      Started server process [11779]
INFO      Waiting for application startup.
INFO      Application startup complete.
INFO      Uvicorn running on http://0.0.0.0:8000 (Press CTRL+C to quit)
```

Figure 13-4. ML Model API running

In Codespaces, you will also see a pop-up as shown in Figure 13-5.

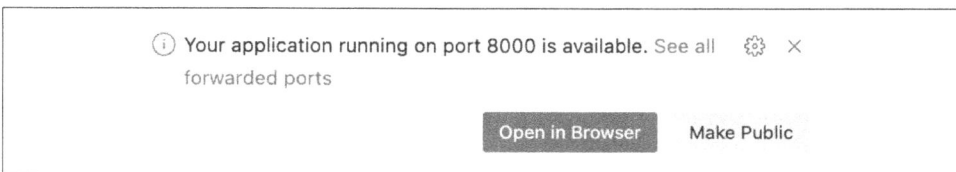

```
ⓘ Your application running on port 8000 is available. See all    ⚙  ✕
  forwarded ports

                                        Open in Browser   Make Public
```

Figure 13-5. Codespaces browser window pop-up

Click "Open in Browser" to open a browser tab outside your Codespaces. This browser will show a base URL ending in *app.github.dev* that contains the response from your API running on Codespaces. You should see the following health check message in your web browser:

```
{"message":"API health check successful"}
```

This confirms your API is running. To view the interactive API documentation for your API, copy and paste the following onto the end of the base URL in your browser: **/docs**. For example, the full URL might be *https://happy-pine-tree-1234-8000.app.github.dev/docs* in the browser. You should see documentation, as shown in Figure 13-6.

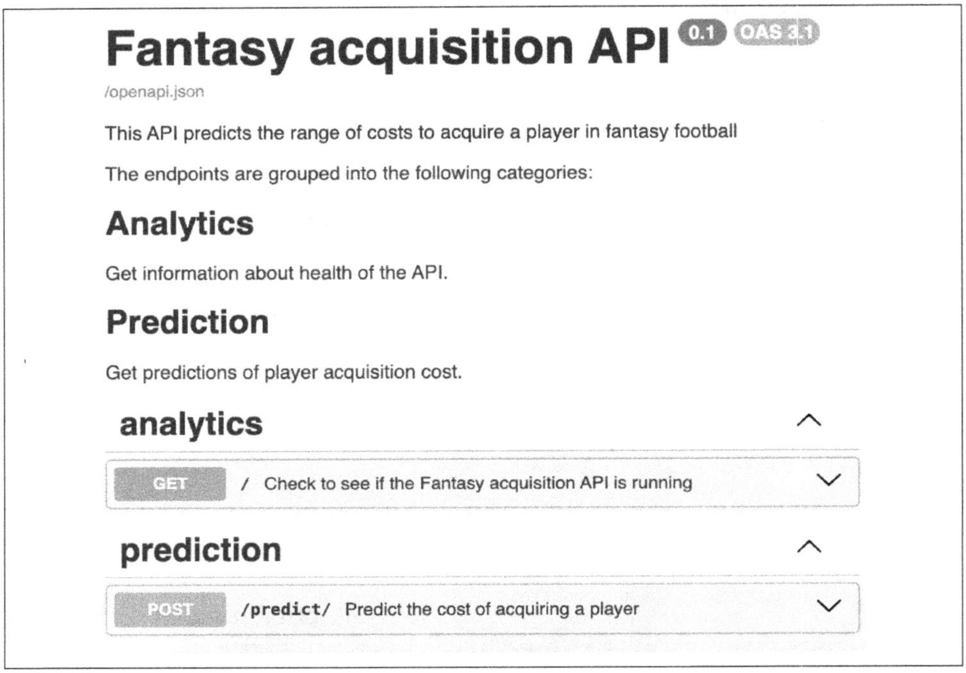

Figure 13-6. Documentation for the ML API

To call the ML API, select the POST */predict/* endpoint to open up that section, and then select "Try it out," after which you should see a display as shown in Figure 13-7.

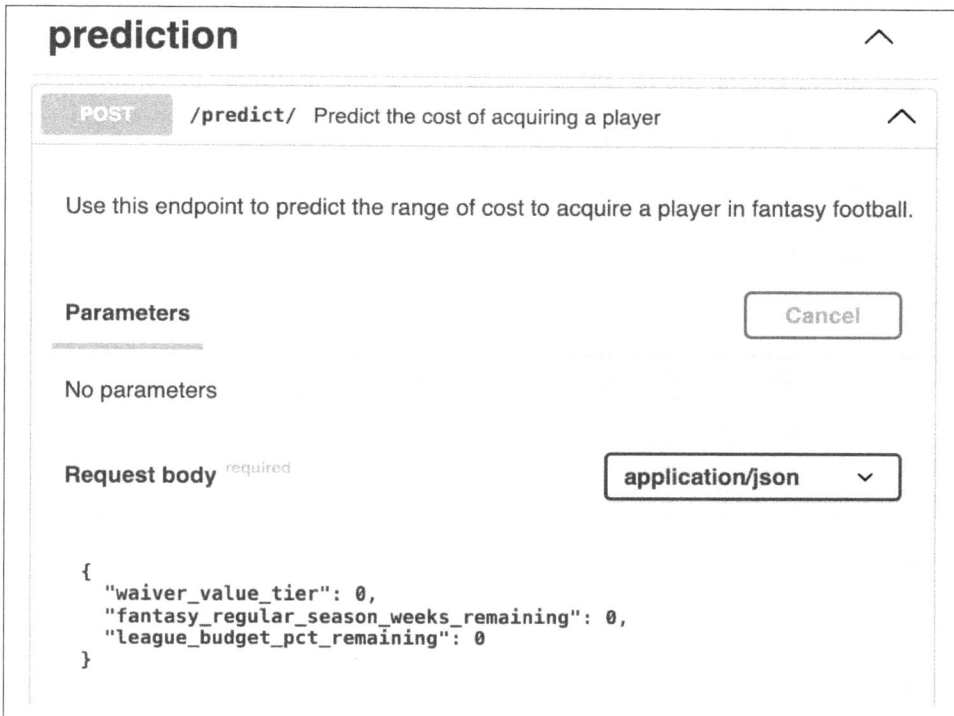

Figure 13-7. Trying it out for /predict

Here is one big difference from the API you created in Part I: you are using a POST endpoint instead of GET. To make a call to a POST endpoint, you provide an HTTP request body in JSON format. This is where users will provide the input values to send to the API. These were automatically generated based on the FantasyAcquisi tionFeatures Pydantic class you defined in the previous section. Update the request body with the following values:

```
{
  "waiver_value_tier": 1,
  "fantasy_regular_season_weeks_remaining": 12,
  "league_budget_pct_remaining": 88
}
```

Click Execute to send these values to the API. Scroll down and you should see a server response with a Code value of 200, which indicates success. The predicted values will be returned in the HTTP response body, which matches the definition of the PredictionOutput Pydantic class. You should see an output similar to Figure 13-8, although the predicted values may differ.

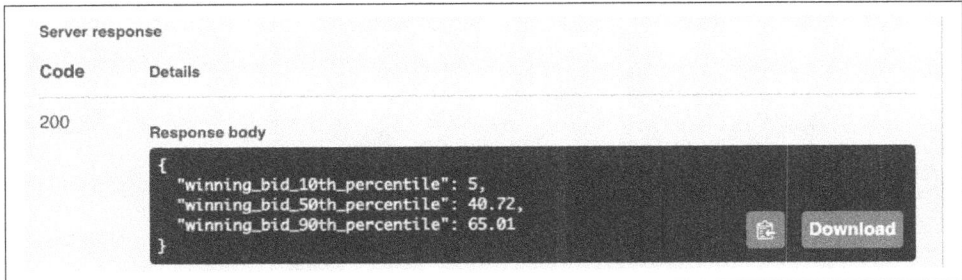

Figure 13-8. API response with prediction

To see the final structure of your project, execute the tree command as follows:

```
.../chapter13 (main) $ tree --prune -I 'build|*.egg-info|__pycache__'
.
├── acquisition_model_10.onnx
├── acquisition_model_50.onnx
├── acquisition_model_90.onnx
├── main.py
├── player_acquisition_model.ipynb
├── player_acquisition_notebook.log
├── player_training_data_full.csv
├── requirements.txt
└── schemas.py

0 directories, 9 files
```

Congratulations! You have created the first draft of an ML model and served it with a REST API.

Documenting Machine Learning Models

Because ML models are not explicitly programmed, the full details of how they operate are not always understood, even by the people who train them. Due to the way the models operate, it is not usually possible to determine why a model makes an individual prediction or generates specific content.

Model providers typically provide documentation on the model to explain its intended purpose, along with known issues and limitations. The techniques for documenting ML models and applications are quickly evolving. Two methods I have come across are model cards and system cards.

Model cards were proposed by Google to be transparent about an ML model's operations, risks and biases. Google's proposal for model cards (*https://oreil.ly/PDB6C*) gives more information about this method, and a few examples.

Meta proposed *system cards* to document a broader AI system that may contain multiple models. Meta's description (*https://oreil.ly/Lx2md*) says system cards look "holistically across an AI system, versus one-off models."

Additional Resources

To learn more about data science projects, read *Practical Data Science with Python* by Nathan George (Packt Publishing, 2021).

To get more experience using scikit-learn and other ML libraries, read *Hands-On Machine Learning with Scikit-Learn, Keras, and Tensorflow* by Aurélien Géron (O'Reilly, 2022).

To learn more about deploying models for prediction, read *Designing Machine Learning Systems* by Chip Huyen (O'Reilly, 2022).

Extending Your Portfolio Project

Now that you have an API template for serving ML models, there are many ways you can extend this project:

- Perform evaluation of this model to determine how effective it is at predictions. Explore tuning parameters and change the features to see if you can improve its effectiveness.
- Create other types of models using scikit-learn and convert them to ONNX format. Use the API to deploy these new models.
- ONNX format also supports libraries other than scikit-learn. Explore creating models with libraries such as PyTorch or XGBoost, convert them to ONNX format, and use the API to deploy these models.

Summary

In this chapter, you learned about the ML lifecycle and created an ML model using scikit-learn. Then, you converted the model to ONNX format to make it compatible with more frameworks. Finally, you deployed your model using FastAPI and used it for real-time inference.

In Chapter 14, you will use generative AI to call an API using LangChain.

Using APIs with LangChain

A system is more "agentic" the more an LLM decides how the system can behave.
—Harrison Chase, LangChain creator

AI applications use LLMs as a natural language interface with users, and researchers are exploring ways to use the LLMs to perform multistep tasks. This chapter examines two important ways that APIs and LLMs are used together to create AI applications. First, you will look at calling LLMs using APIs, and then you will reverse the roles and call APIs with LLMs. You will use LangChain for both of these tasks.

LangChain and its related project, LangGraph, are open source frameworks for creating *agentic* applications—applications that use LLMs to control the system behavior. Although many developers build these applications by calling the LLM APIs directly and performing custom coding to interact with them, LangChain and LangGraph standardize many of the tasks required. You can think of them as frameworks that sit on top of the APIs or models.

Here are a few new terms:

Agent
> Harrison Chase defines an agent as "a system that uses an LLM to decide the control flow of an application." Agents are not preprogrammed, like traditional software—they use a model to reason and decide the flow of a conversation. They can execute tool calls that are suggested by function-calling models.

Function-calling model
> This is a specialized type of model that considers available *functions* or *tools* and suggests when they should be used. Despite the name, the models don't call the tool directly; they give that suggestion to *agents*, who do the calling.

Models
> These are the AI models that LangChain users call. LangChain can use models that are downloaded locally or called via web APIs provided by model providers.

Model families
> These are multiple models that share a name and architecture.

Toolkit
> This is a collection of multiple tools that an agent will use to perform tasks.

Tools or functions
> This is code that provides extra skills to an agent. A simple Python function might perform mathematical operations. You will create a tool that calls the SWC API.

The software introduced in this chapter will focus on using the LangChain ecosystem with APIs. Table 14-1 displays the new tools you will use.

Table 14-1. Tools used in this chapter

Software name	Version	Purpose
LangChain	0.3	Python library used to create tools and toolkits that allow agents to use your API
LangGraph	0.2	Python library used to create an agent
Anthropic Claude	3.5 Sonnet	Model used to provide reasoning to the LangGraph agent
Pydantic	2	Python library used to perform validation in your toolkit
swcpy	NA	SDK created in Chapter 7 for your API

Calling AI Using APIs (via LangChain)

Figure 14-1 shows the high-level architecture of the project you will create in this chapter.

To understand the capabilities that LangChain is harnessing, it is helpful to look at the APIs that LangChain interacts with. The LangChain "Chat Models" page (*https://oreil.ly/2elFn*) lists more than 60 providers of chat models. The list also displays which models are function-calling, which will be important in this chapter because they are capable of using tools to call APIs.

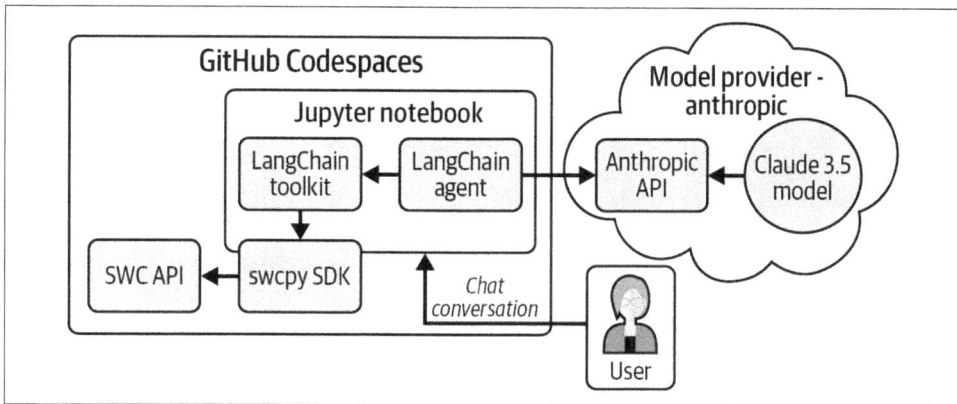

Figure 14-1. High-level architecture

If you look at the documentation for these model providers, they typically provide documentation for their APIs and allow you to register for an API key (for a fee). These companies follow many of the best practices from Chapter 6: interactive API documentation, SDKs in Python and other languages, methods of contacting support, and many other features. For these providers, APIs are not a side business—they *are* the business.

> While it's tempting to focus only on the potential of LLMs, the model providers also publish some sobering warnings. The providers often provide a *model card* or *system card* that describes the intended use and limitations of their models. At the time of this writing, these cards of major LLMs list risks such as bias, hallucinations, mistakes, and harmful content. For example, Anthropic states in their article, "The Claude 3 Model Family: Opus, Sonnet, Haiku": "The models should not be used on their own in high-stakes situations" (*https://oreil.ly/vq7FK*).
>
> In this chapter, you will be using a read-only fantasy football API, which is low risk.

Creating a LangGraph Agent

Letting an LLM decide the control flow of an application (i.e. what we call agents) is attractive, as they can unlock a variety of tasks that couldn't previously be automated. In practice, however, it is incredibly difficult to build systems that reliably execute on these tasks.

—LangChain blog post, "Announcing LangGraph v0.1 & LangGraph Cloud: Running agents at scale, reliably" (*https://oreil.ly/KoxAw*), June 2024

LangGraph is a LangChain-related project focused on creating applications that have one or more agents working together. LangChain has methods for using agents, which have recently been labeled as *legacy methods*, at the time of this writing. Lang-Graph is intended to be an improvement on these methods by allowing more developer control and supporting multiagent applications.

LangGraph uses some terminology from mathematical graph theory (see "Orchestrating the Data Pipeline with Apache Airflow" on page 199). The logical flow of LangGraph agents is represented by *nodes*, which are processes that update the *state* of the application. The nodes are connected by *edges*, which are one-way flows between one node and another. Where Airflow had *acyclic* graphs that begin at a start node and go in one direction to an end node, LangGraph allows *cyclical* graphs, where nodes and edges can loop multiple times. For example, a loop may allow a user to ask multiple questions to a model before getting the final answer.

Figure 14-2 shows how nodes and edges relate to each other in a *directed cyclic graph* that contains loops.

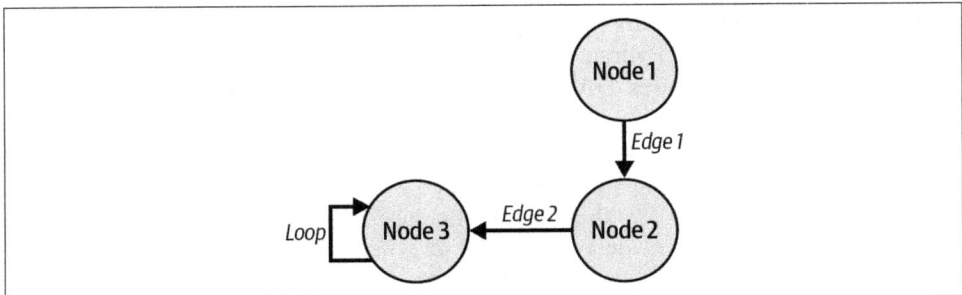

Figure 14-2. Directed cyclic graph

You will be creating an agent in this chapter that can be represented by a directed cyclic graph.

Signing Up for Anthropic

You will be using a model from Anthropic, so you need to sign up for an Anthropic account at the "Build with Claude" web page (*https://console.anthropic.com/login*).

To use the Anthropic API, you will need to upgrade your initial account to a Build Plan and add funds to your account. To limit potential overspending, I suggest initially adding $5 and disabling auto-reload. This means the maximum amount you can spend is $5, even if your API key is misused.

Next, navigate to the Anthropic API keys page (*https://oreil.ly/fcaYA*) and select Create Key. Give the key a name such as **secret-api-key** and select the Default workspace. Your display should look like Figure 14-3.

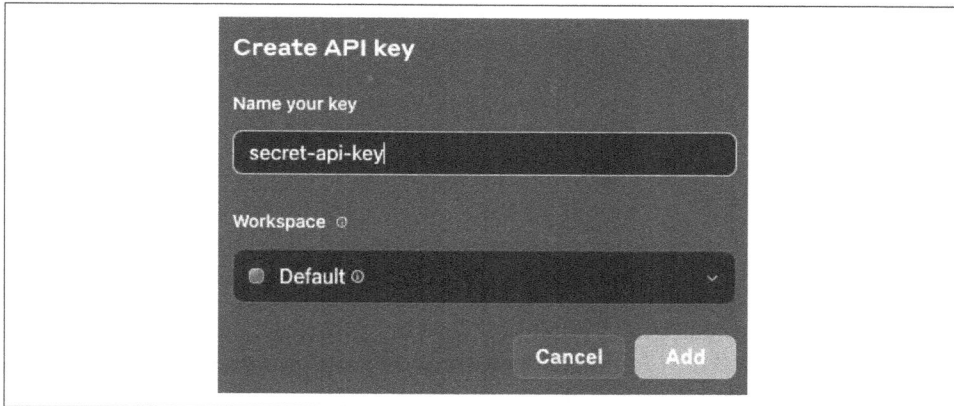

Figure 14-3. Creating an Anthropic API key

Click Add. You will be prompted to "Save Your API key." This is the only time the key will be displayed. Store this key in a secure location.

> If you lose your Anthropic API key or think it may have been exposed, delete it from the keys page and create a new one. This is called *rotating* a credential, and it is a convenient way to lower the risk of your API key being abused.

Launching Your GitHub Codespace

Before you launch the GitHub Codespace for this chapter, you want to add the Anthropic API key as a secret in your Codespace. This will allow you to use the Anthropic models with LangGraph.

Follow the instructions for "Adding a secret" (*https://oreil.ly/VGSJ3*) and create a new secret with the following information:

- *Name*: ANTHROPIC_API_KEY
- *Value*: Use the value of the Anthropic API key you created in the previous section.
- *Repository access:* Select the ai-project repository.

When completed, you should see the key saved as shown in Figure 14-4.

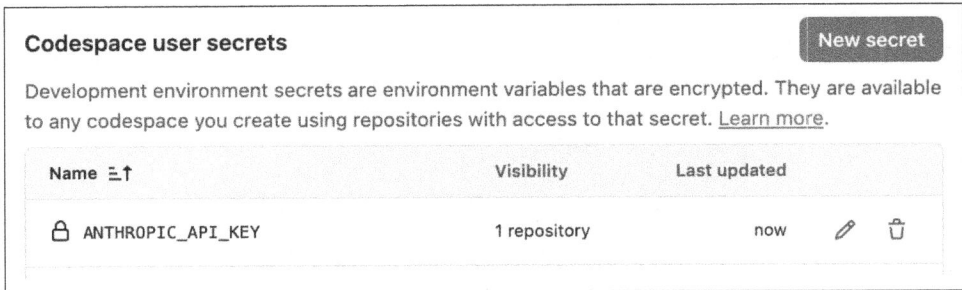

Figure 14-4. Codespaces secrets

Installing the New Libraries in Your Codespace

Now launch your Codespace. You will be working with the Part III GitHub Codespace that you created in "Cloning the Part III Repository" on page 240. If you haven't created your Part III Codespace yet, you can complete that section now. To install the libraries you need for this chapter, create a file named *chapter14/requirements.txt*:

```
.../analytics-project (main) $ cd chapter14
.../chapter14 (main) $ touch requirements.txt
```

Update *chapter14/requirements.txt* with the following contents:

```
langchain_core>=0.3.0,<0.4.0 ❶
langchain_anthropic>=0.2.0,<0.3.0 ❷
langgraph>=0.2.0,<0.3.0   ❸
```

❶ This installs the base LangChain library and its dependencies.

❷ This library is used for the specific model you use in this chapter.

❸ This is the LangGraph library used to define your agent.

Execute the following command to install the new libraries in your Codespace, and make sure to upgrade to the latest version available:

```
.../chapter14 (main) $ pip3 install --upgrade -r requirements.txt
```

You should see a message that states that these libraries were successfully installed.

Creating Your Jupyter Notebook

You will be using a Jupyter Notebook to implement a conversational chat agent, based on an example from the LangChain GitHub repository (*https://oreil.ly/U-TJF*).

To get started, run the following commands in the Terminal window to create the new directory and the Jupyter Notebook you will be using in this chapter:

```
.../chapter14 (main) $ touch langgraph_notebook.ipynb
```

Click the *langgraph_notebook.ipynb* file to open it. In the top-right of the file, click Select Kernel. Codespaces should prompt you to "Install/Enable suggested extensions Python + Jupyter"—select this. Click Install in the additional pop-up window, if prompted. This will install some extensions to VS Code.

After the installation completes, the title of the window will change to Select Another Kernel. Select Python Environments. The title of the window will change to Select a Python Environment. One Python version should be listed with a star next to it—select this Python version.

Click "+ Markdown" to create a new Markdown cell. Enter the following title in the Markdown cell:

```
# LangGraph Agent
## Without tools
```

Run this cell by clicking the play icon on the left of the cell or by pressing Shift-Enter. You should see your message formatted as a title.

Create another Markdown cell and run it with the following:

```
## Library Imports
```

Hover your cursor below this cell and click "+ Code" to create a new Python cell. Enter the following code in the Python cell:

```
from langchain_core.messages import HumanMessage ❶
from langchain_anthropic import ChatAnthropic ❷
from langgraph.checkpoint.memory import MemorySaver ❸
from langgraph.graph import END, START, StateGraph, MessagesState
import logging
from IPython.display import Image, Markdown, display ❹
from langchain_core.runnables.graph import CurveStyle,
MermaidDrawMethod, NodeStyles ❺
```

❶ This import defines the messages you'll use to communicate with the agent.

❷ This is the library for using the Anthropic model.

❸ The following libraries enable different parts of the LangGraph framework.

❹ This will be used for the formatting of the messages returned by the agent.

❺ This will be used to display a visual view of the graph.

Placing all the imports at the top of your notebook helps keep track of the libraries you are using. These imports will work for all the cells in this Jupyter Notebook.

Add another Markdown cell with the following text:

```
## Configure logging
```

Add and run a Python code cell with the following:

```
for handler in logging.root.handlers[:]: ❶
    logging.root.removeHandler(handler)

logging.basicConfig(
    filename='langgraph_notebook.log',
    level=logging.INFO,  ❷
)
```

❶ This statement removes any existing logging handlers configured by Codespaces.

❷ This sets the logging level to record in the log. Review Table 7-1 for more details about Python logging.

Add another Markdown cell with the following text:

```
## Configure Agent and Model
```

Add and run a Python code cell with the following:

```
model = ChatAnthropic(model="claude-3-5-sonnet-20240620", temperature=0) ❶

def call_model(state: MessagesState): ❷
    messages = state['messages']
    response = model.invoke(messages)
    return {"messages": [response]}

workflow = StateGraph(MessagesState) ❸

workflow.add_node("agent", call_model) ❹

workflow.add_edge(START, "agent") ❺

checkpointer = MemorySaver()

app = workflow.compile(checkpointer=checkpointer) ❻
```

❶ This initializes the Anthropic model. The `temperature` sets how creative the responses are. Lower is less creative and more predictable.

❷ This function is used to send messages to the model using the `state` object.

❸ The `workflow` is used to define the tasks available to the agent.

❹ The `agent` node will use the `call_model` function defined above.

❺ This defines the start of the graph, which will directly call the `agent` node.

❻ This statement compiles the graph into a LangChain `Runnable` object named `app`. (For more information, see the documentation for the Runnable interface (*https://oreil.ly/UJJzq*).) The `app` object will act as the agent.

Viewing a visual representation of the graph will help you see the nodes and edges that were created in this code, and the flow of the agent you created. Add another Markdown cell with the following text:

```
## Visualize the Graph
```

Add and run a Python code cell with the following:

```
display(
    Image(
        app.get_graph().draw_mermaid_png(
            draw_method=MermaidDrawMethod.API,
        )
    )
)
```

Displaying an image of the graph demonstrates the flow of your agent. You should see an image that matches Figure 14-5. This shows that the graph will start and a message will be sent to the agent without referencing any other nodes.

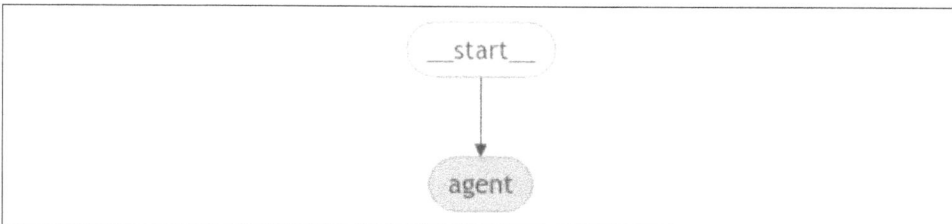

Figure 14-5. Graph of the basic agent

Chatting with the LangGraph Agent

You are ready to have a conversation with the agent you have created. You will accomplish this by using the `app.invoke()` method along with a `messages` object. The call to the `invoke()` method also includes a `config` object with a `thread_id` value. The `thread_id` allows the agent to remember previous messages in the conversation.

Add another Markdown cell with the following text:

```
## Chat with the Agent
```

Add and run a Python code cell with the following:

```
final_state = app.invoke(
    {"messages": [HumanMessage(content="What teams did Joe Montana play for?")]},
    config={"configurable": {"thread_id": 99}}
)
display(Markdown(final_state["messages"][-1].content))
```

This is how you will chat with your agent. The `app.invoke` command sends the `messages` object and adds a question as a `HumanMessage`. The `config` object passes a `thread_id` so that the agent knows you are in the same conversation. While you are using this `thread_id`, the agent will remember the previous messages in the same conversation. After each `invoke()` statement, the `final_state` object contains a list of messages between the model and the human.

To get the most recent message in the conversation, you will display `final_state["messages"][-1].content`. The messages will have Markdown formatting, so you can use the `display(Markdown())` statement to format them correctly.

Remember that models are *nondeterministic*, even with a temperature setting of zero. This means that each interaction may be slightly different. But if your call to the graph is successful, you should see a message similar to this:

```
Joe Montana played for two NFL teams during his professional career:

1. San Francisco 49ers (1979-1992)
2. Kansas City Chiefs (1993-1994)
He spent the majority of his career with the 49ers, where he achieved his
greatest success, before finishing his career with two seasons in Kansas City.
```

> If you receive an error at this step related to the API key, check to make sure that you have added funds to your Anthropic account, created an API key, and set it in your Codespaces secrets. If any of these steps were skipped, you won't be able to use the Anthropic model in this example.

It appears that the Anthropic model has been trained with historical information about the NFL. Let's see how it does when we ask questions about the SportsWorld-Central app. Add and run a Python code cell with the following:

```
final_state = app.invoke(
    {"messages": [HumanMessage(content="What are the leagues in the
    SportsWorldCentral fantasy football platform?")]},
    config={"configurable": {"thread_id": 99}}
)
display(Markdown(final_state["messages"][-1].content))
```

You will receive a message similar to the following:

```
I apologize, but I don't have any specific information about leagues in a
platform called "SportsWorldCentral" for fantasy football. Fantasy football
platforms can vary widely in their league structures and offerings, and I'm not
familiar with this particular one. If this is a real platform, you might
need to check their website or contact their customer support
for accurate information about their league types and structures.
If you have more general questions about fantasy football leagues,
I'd be happy to help with those.
```

The model apparently suspects that SportsWorldCentral may not be "a real platform," but even for real-world platforms like Sleeper or MyFantasyLeague.com, the model doesn't contain information about specific leagues and teams.

To solve this knowledge gap, you will create a toolkit that provides access to the SportsWorldCentral API.

Running the SportsWorldCentral (SWC) API Locally

The code and database files for the SWC API are in the */api* folder of your Codespace. Start a second session in the Terminal and install the required libraries for the API in your Codespace as shown, using the *requirements.txt* file that is provided:

```
.../ai-project (main) $ cd api
.../api (main) $ pip3 install -r requirements.txt
```

Now launch the API from the command line as shown:

```
.../api (main) $ fastapi run main.py
```

You will see several messages from the FastAPI CLI, ending with the following:

```
INFO:     Started server process [19192]
INFO:     Waiting for application startup.
INFO:     Application startup complete.
INFO:     Uvicorn running on http://0.0.0.0:8000 (Press CTRL+C to quit)
```

The API is now running in Codespaces. Copy the URL shown in the Terminal (in this example it is *http://0.0.0.0:8000*, and it may be in yours too).

In the primary Terminal session, create a file named *.env* in the base project directory as shown:

```
chapter14 (main) $ touch ../.env
```

Update this file with the following contents:

```
SWC_API_BASE_URL=[URL from previous step]
```

This will be used in the Jupyter Notebook you create later.

Installing the swcpy Software Development Kit (SDK)

To build the toolkit for using your API, you will first install the swcpy SDK that you created in Chapter 7. If you completed that project already, you can copy that code to this repository and install it locally using `pip3 install -e` in the */sdk* directory. If you have not completed that yet, execute the following command to install the SDK from the Part I GitHub repository:

```
.../api (main) $ pip install swcpy@git+https://github.com/handsonapibook/
api-book-part-one#subdirectory=chapter7/complete/sdk
```

If the installation is successful, you should see the following:

```
Successfully built swcpy
Installing collected packages: pyarrow, backoff, swcpy
Successfully installed backoff-2.2.1 pyarrow-16.1.0 swcpy-0.0.2
```

> You will be creating a tool that uses the swcpy SDK, which gives the most robust way to interact with your API. There are other methods you could use to interact with your API. For example, you could also build a tool that is based on the simple API client that you created in Chapter 9. Or you could create a tool that directly calls individual API endpoints using httpx or requests. You may want to experiment with different methods to see which you prefer.

Creating a LangChain Toolkit

A LangChain toolkit contains multiple tools. LangChain provides multiple ways to create tools, as shown in "How to Create Tools" (*https://oreil.ly/V8tBi*). You will create tools by subclassing the `BaseTool` class. According to the documentation, this method "provides maximal control over the tool definition."

Create a Python file named *swc_toolkit.py* as shown:

```
chapter14 (main) $ touch swc_toolkit.py
```

At the top of your toolkit file, you will add your imports and load the Codespace secret. Add the following contents to *swc_toolkit.py*:

```
import os
from typing import Optional, Type, List

from pydantic import BaseModel, Field

from langchain_core.callbacks import CallbackManagerForToolRun ❶
from langchain_core.tools import BaseTool, BaseToolkit

try: ❷
        from swcpy import SWCClient
```

```
            from swcpy import SWCConfig
            from swcpy.swc_client import League, Team
    except ImportError:
        raise ImportError(
            "swcpy is not installed. Please install it."
        )

    config = SWCConfig(backoff=False) ❸
    local_swc_client = SWCClient(config)
```

❶ These imports use the LangChain libraries for defining tools and toolkits.

❷ This statement checks to see if the swcpy SDK has been installed in the environment. If not, an error is generated.

❸ This instantiates the SDK. For more information about how this works, see Chapter 7.

The swcpy SDK has multiple functions, which call API endpoints in different ways. For each of the SDK functions you make available to the agent, create an instance of the Pydantic BasesModel class for input values and an instance of the LangChain BaseTool class to call the SDK. You will create the following tools:

- HealthCheckTool: Allows the agent to check if the API is up and running
- ListLeaguesTool: Allows the agent to get a list of leagues
- ListTeamsTool: Allows the agent to get a list of teams

Add the following contents to the bottom of *swc_toolkit.py*, which add the Health CheckTool and ListLeaguesTool:

```
class HealthCheckInput(BaseModel): ❶
    pass

class HealthCheckTool(BaseTool): ❷
    name: str = "HealthCheck"
    description: str = (
        "useful to check if the API is running before you make other calls"
    args_schema: Type[HealthCheckInput] = HealthCheckInput ❸
    return_direct: bool = False

    def _run(
        self, run_manager: Optional[CallbackManagerForToolRun] = None
    ) -> str:
        """Use the tool to check if the API is running."""
        health_check_response = local_swc_client.get_health_check() ❹
        return health_check_response.text
```

```
class LeaguesInput(BaseModel):  ❺
    league_name: Optional[str] = Field(
        default=None,
        description="league name. Leave blank or None to get all leagues.")

class ListLeaguesTool(BaseTool):  ❻
    name: str = "ListLeagues"
    description: str = (
        "get a list of leagues from SportsWorldCentral. "
        "Leagues contain teams if they are present."
    args_schema: Type[LeaguesInput] = LeaguesInput
    return_direct: bool = False

    def _run(
        self, league_name: Optional[str] = None,
        run_manager: Optional[CallbackManagerForToolRun] = None
    ) -> List[League]:
        """Use the tool to get a list of leagues from SportsWorldCentral."""
        # Call the API with league_name, which could be None
        list_leagues_response = local_swc_client.list_leagues(
                                league_name=league_name)
        return list_leagues_response
```

❶ For each tool, you will define an input object, which is a subclass of the Pydantic
   BaseModel. This tool does not accept parameters, so you are creating an empty
   object.

❷ This is the definition of the tool for the health check, and it is a subclass of the
   LangChain BaseTool. The information you provide in this section will be used by
   the model to decide how and when to use this tool.

❸ The args_schema defines what inputs are expected, and the model will use this to
   send input to the tool.

❹ This is where the action occurs. It is calling the get_health_check() method
   from the SDK.

❺ This is the input defined for the ListLeaguesTool. It contains parameters used
   by that tool.

❻ This is the tool that will call the SDK's list_leagues() function.

Now you will add the ListTeamsTool and create a BaseToolkit object that represents
all of the tools that you will be providing to an agent. Add the following contents to
the bottom of *swc_toolkit.py*:

```
class TeamsInput(BaseModel):  ❶
    team_name: Optional[str] = Field(
```

```
            default=None,
            description="Name of the team to search for.
        Leave blank or None to get all teams.")
    league_id: Optional[int] = Field(
            default=None,
            description=(
                "League ID from a league. You must provide a numerical League ID. "
                "Leave blank or None to get teams from all leagues."
                ))


class ListTeamsTool(BaseTool):  ❷
    name: str = "ListTeams"
    description: str = (
            "Get a list of teams from SportsWorldCentral. Teams contain players "
            "if they are present. Optionally provide a numerical League ID to "
            "filter teams from a specific league.")
    args_schema: Type[TeamsInput] = TeamsInput
    return_direct: bool = False


    def _run(
        self, team_name: Optional[str] = None,
        league_id: Optional[int] = None,
        run_manager: Optional[CallbackManagerForToolRun] = None
    ) -> List[Team]:
        """Use the tool to get a list of teams from SportsWorldCentral."""
        list_teams_response = local_swc_client.list_teams(
            team_name=team_name, league_id= league_id)
        return list_teams_response

class SportsWorldCentralToolkit(BaseToolkit):  ❸
    def get_tools(self) -> List[BaseTool]:  ❹
        """Return the list of tools in the toolkit."""
        return [HealthCheckTool(), ListLeaguesTool(), ListTeamsTool()]  ❺
```

❶ This is the input defined for the ListTeamsTool. It contains parameters used by that tool.

❷ This is the tool that will call the SDK's list_teams() function.

❸ The toolkit is subclassed from the LangChain BaseToolkit class.

❹ The get_tools() method returns a list of the tools in the toolkit.

❺ This statement instantiates the tools and returns them in a list.

Calling APIs Using AI (with LangGraph)

In the *langgraph_notebook.ipynb* notebook, you created an agent that could chat, but the agent was unable to answer questions from the SWC API. Now you will make an improved version that can use tools to call the SWC API. First, make a copy of the Jupyter Notebook with the following commands:

```
.../chapter14 (main) $ cp langgraph_notebook.ipynb \
langgraph_notebook_with_toolkit.ipynb
```

Open *langgraph_notebook_with_toolkit.ipynb* and select the Python kernel as you did previously. Update the first Markdown cell with the following text for clarity:

```
# LangGraph Agent
## _With_ tools
```

Add the following to the first Python cell that contains the `import` statements and re-run it:

```
from swc_toolkit import SportsWorldCentralToolkit ❶
from langchain_core.tools import tool ❷
from langgraph.prebuilt import ToolNode ❸
from typing import Literal
```

❶ This is the `import` statement for the toolkit that you created.

❷ This `import` provides the LangChain tool functionality.

❸ This `import` adds LangGraph support for toolkits.

Find the Python cell that is used to configure logging. Beneath that cell, add the following Markdown cell:

```
## Create toolkit
```

Now you will reference the toolkit you created and imported earlier. To instantiate your toolkit, add the following Python cell and run it:

```
swc_toolkit = SportsWorldCentralToolkit()
tools = swc_toolkit.get_tools()
```

These statements create an instance of your toolkit class, then retrieve all of the tools from it and put those into the `tools` object.

For the main body of your agent, there will be many changes. You will be *binding* the tools to your model, which means making the model aware of the tools it has access to. Then, you will create a node for the tools and update the flow of the agent to use tools when needed to answer questions.

Locate the Python cell following the title "Configure Agent and Model." Replace the entire contents with the following, and run it:

```
tool_node = ToolNode(tools)

model = ChatAnthropic(model="claude-3-5-sonnet-20240620",
                      temperature=0).bind_tools(tools) ❶

def should_continue(state: MessagesState) -> Literal["tools", END]: ❷
    messages = state['messages']
    last_message = messages[-1]
    if last_message.tool_calls:
        return "tools"
    return END

def call_model(state: MessagesState):
    messages = state['messages']
    response = model.invoke(messages)
    return {"messages": [response]}

workflow = StateGraph(MessagesState)

workflow.add_node("agent", call_model)
workflow.add_node("tools", tool_node) ❸

workflow.add_edge(START, "agent")

workflow.add_conditional_edges( ❹
    "agent",
    should_continue,
)

workflow.add_edge("tools", 'agent')

checkpointer = MemorySaver()

app = workflow.compile(checkpointer=checkpointer)
```

❶ This adds the `bind_tools()` function to the model initialization so that the model can use the toolkit you created.

❷ This will be used as a *conditional edge*, which is a workflow that decides how to proceed based on the results. If the model suggests a tool call, this returns the literal "tools."

❸ This adds a new node in the graph to perform tool calls using the new toolkit.

❹ This adds the conditional edge that uses the `should_continue` function defined earlier.

To see how the flow has been updated to add a tool node, rerun the Python cell under the "Visualize the Graph" heading.

You should see an image that matches Figure 14-6. This shows a loop between the agent and the available tools.

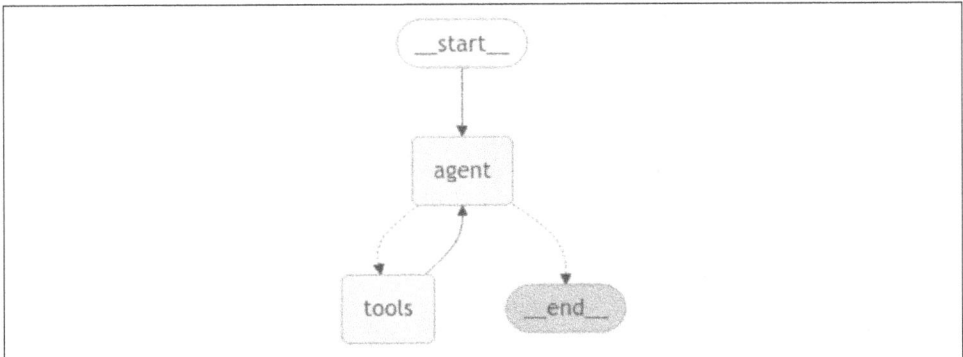

Figure 14-6. Graph of the agent with tools

The agent will decide how many times it needs to use the tools to answer the question. Depending on the question asked and the reasoning of the agent, this may involve one API call, multiple API calls, or no API calls.

Chatting with Your Agent (with Tools)

Now it's time to see if your effort to provide tools to the agent has improved its ability to answer questions about SportsWorldCentral. Let's start with the question that stumped it before. Enter the following in a new Python cell and run it:

```
final_state = app.invoke(
    {"messages": [HumanMessage(content="What are the leagues in the
    SportsWorldCentral fantasy football platform? Keep the response simple.")]},
    config={"configurable": {"thread_id": 99}}
)
display(Markdown(final_state["messages"][-1].content))
```

You will receive a message similar to the following:

```
Here's a simple list of the leagues in the SportsWorldCentral fantasy
football platform:

1. Pigskin Prodigal Fantasy League
2. Recurring Champions League
3. AHAHFZZFFFL
4. Gridiron Gurus Fantasy League
5. Best League Ever

These are the five leagues currently available in the platform.
```

If you look at the Terminal window, you should see the following, which indicates that the agent used the `ListLeaguesTool` from the toolkit:

```
INFO: 0.0.0.0:8000 - "GET /v0/leagues/?skip=0&limit=100 HTTP/1.1" 200 OK
```

Congratulations! You have created a LangGraph agent and provided it a toolkit to call your custom API.

Take time and experiment with questions that will encourage the agent to use the `HealthCheckTool` and the `ListTeamsTool`. Look at the Terminal output or log file to see the API calls that are generated. How can you influence its tool use? How can you guide it to answer more difficult questions using the tools? How can you stump it?

Extending Your Portfolio Project

You have learned to create AI agents using LangGraph and LangChain. Here are a few ideas to continue your learning:

- If you deployed your API in Chapter 6, modify the code in this chapter to use the cloud-hosted API instead of running the API in your Codespace.

- Continue adding more tools from your API to the toolkit. Explore how the definition of the tools and the prompts you provide influence the behavior of the agent.

- You used models from Anthropic in this chapter. Experiment using other function-calling models that LangChain supports, as shown in the LangChain "Chat Models" page (*https://oreil.ly/w8YWi*).

- Provide your agent with additional tools and ask it to use these tools in combination. A first suggestion: create a toolkit for the nfl_data_py library you used in "Installing Streamlit and nfl_data_py" on page 217.

Additional Resources

LangChain is a recent framework, so published resources are having a hard time keeping up. As you look at online resources and demos, check the versions of libraries that are being discussed, because at the time of this writing, some of the resources contained deprecated code.

With that caveat, the most up-to-date resource is the official LangChain docs page (*https://oreil.ly/jyC-I*).

LangGraph is even newer than LangChain; the LangGraph documentation (*https://oreil.ly/WBOXF*) has the most detailed information.

To see Anthropic's advice on agents, including several recommended workflows, read "Building Effective Agents" (*https://oreil.ly/QvAKC*).

Google produced an Agents Whitepaper (*https://oreil.ly/M5hV_*) that explains the key components in a fully functional agent. It has a strong emphasis on how agents can use tools, which includes APIs.

Summary

In this chapter, you used the LangChain and LangGraph frameworks to create an intelligent agent that interacted with an AI model and your SWC API. This allowed you to ask questions about SWC data with natural language in a Jupyter Notebook. These are the beginning steps to using APIs with AI applications.

In Chapter 15, you will use ChatGPT, which is a full-featured application that uses LLMs and can call APIs like you did in this chapter. You will create a custom action and custom GPT that can interact with the SWC API.

Using ChatGPT to Call Your API

In the previous chapter, you built a basic generative AI application that could chat with a model in natural language and retrieve data from the SportsWorldCentral API. You created quite a bit of Python code to accomplish that.

In this chapter, you will use custom GPTs from OpenAI to accomplish this task without creating any Python code other than the SportsWorldCentral API. You can think of a custom GPT as a low-code method for creating a generative AI application that connects to your API.

Architecture of Your Application

Figure 15-1 shows the high-level architecture of the project you will create in this chapter.

If you compare this diagram to Figure 14-1, you will see a few similarities. In both cases, a user is using natural language chat to retrieve data from the SportsWorldCentral API. In both cases, a function-calling LLM is used to chat with the user and decide when to call the API for additional information. Although you used an Anthropic model in Chapter 14, you could have used the same OpenAI GPT-4o model that you will use in this chapter.

However, there are also large differences from Chapter 14's architecture. Where Chapter 14 required many different Python components to be developed and run on GitHub Codespaces, in this chapter, only the SportsWorldCentral API will be running there. As you continue through this chapter, I will share more contrasts with Chapter 14.

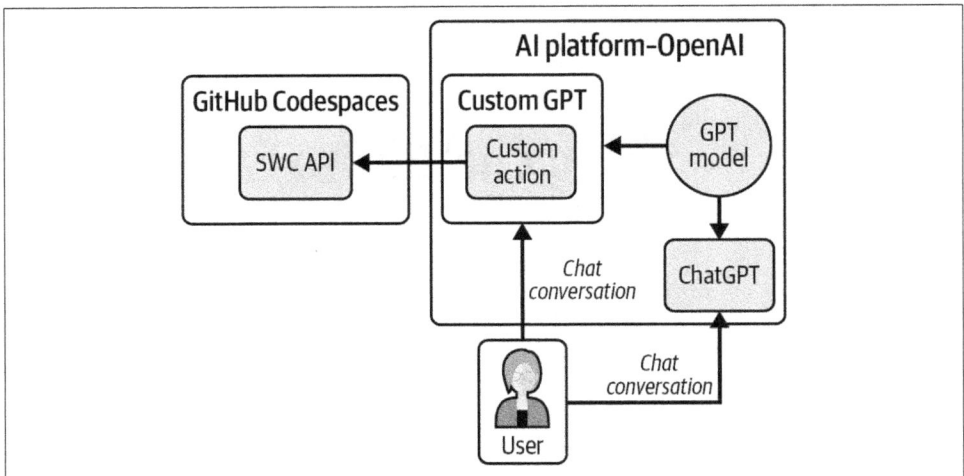

Figure 15-1. High-level architecture

> As with the model used in Chapter 14, you should be aware of the limitations and risks of the model you are using. The GPT-4o System Card (*https://oreil.ly/v8SNX*) states risks that include misinformation, violent speech, and others.

Getting Started with ChatGPT

The first step is to sign up for a ChatGPT user account, from the OpenAI home page (*https://chat.openai.com*). You need to select a plan that includes the Create GPTs capability. At the time of this writing, the ChatGPT pricing page (*https://oreil.ly/gptpri*) shows that Plus, Team, and Enterprise accounts have that option. Follow the sign-up instructions to activate your account. Once you have signed up, log in and navigate back to the ChatGPT home page (*https://chat.openai.com*).

> If you would like to prevent OpenAI from using your chats to train its models, you can request this through the OpenAI privacy portal (*https://oreil.ly/2qLwN*). Doing so reduces the possibility that information you enter in ChatGPT or custom GPTs will be shown to other users accidentally.

Creating a Custom GPT

To create your first GPT, click your profile photo and select My GPTs. You will see the My GPTs page, as shown in Figure 15-2.

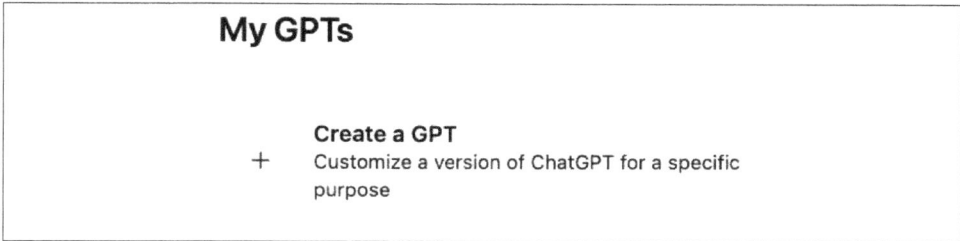

My GPTs

      Create a GPT
+   Customize a version of ChatGPT for a specific
      purpose

Figure 15-2. My GPTs page

Click Create a GPT. The New GPT page will be displayed with the Configure option selected, as shown in Figure 15-3.

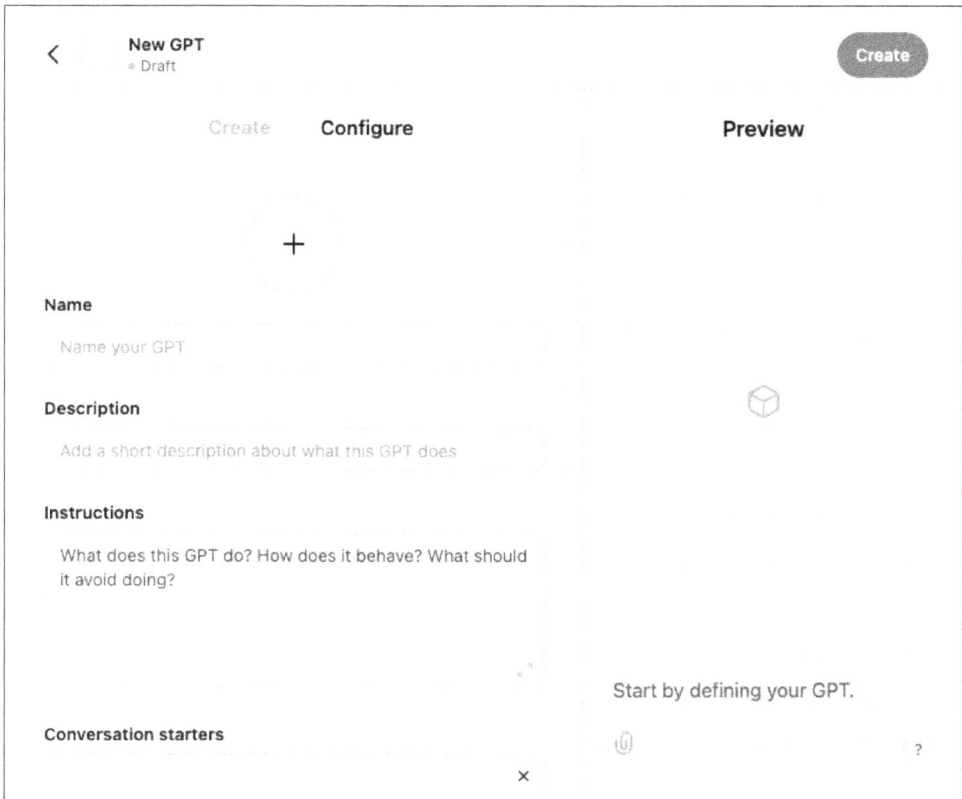

New GPT
● Draft

Create   **Configure**   **Preview**

+

Name

Name your GPT

Description

Add a short description about what this GPT does

Instructions

What does this GPT do? How does it behave? What should it avoid doing?

Start by defining your GPT.

Conversation starters

Figure 15-3. Creating a new GPT

The large button with the plus sign on it is used to select an image for your custom GPT. Download a public-domain image to your computer, then click that button. Select Upload Photo and upload the local file. You can view an example of a public domain image of a football and helmet at *https://oreil.ly/dItdO*.

To configure the GPT, fill out the options on the New GPT page as shown in Figure 15-4.

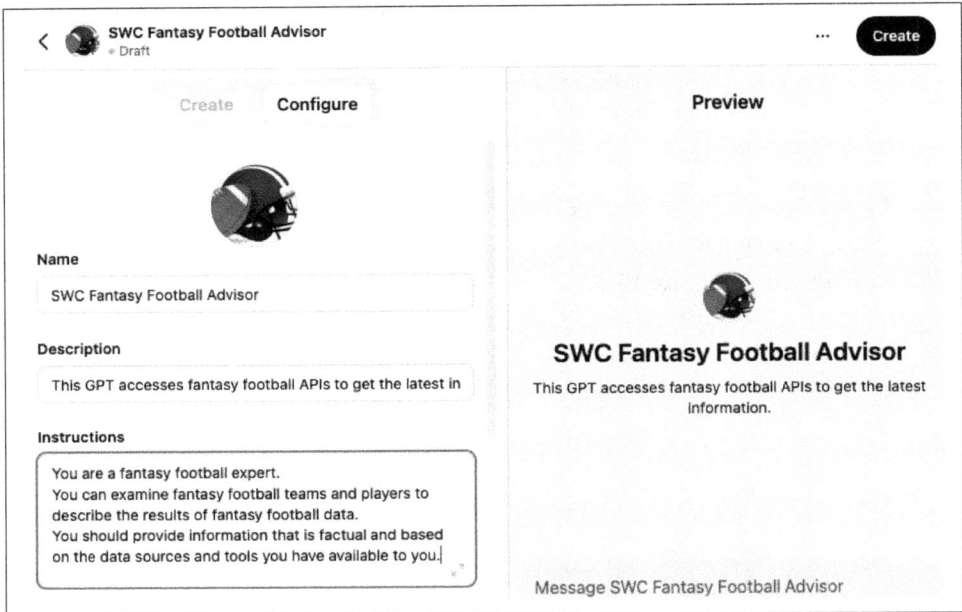

Figure 15-4. New GPT information, part 1

Scroll down the page to see additional options. Leave the conversation starters empty at this time. Under the Capabilities section, do not select any capabilities. The display should look like Figure 15-5.

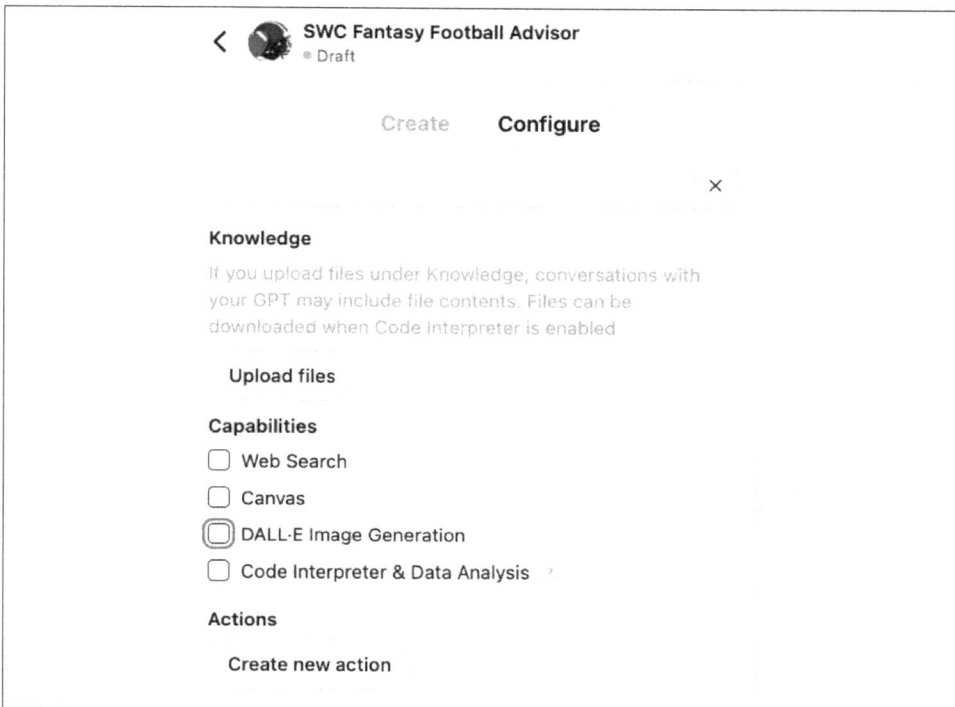

Figure 15-5. New GPT information, part 2

Take a minute to see what your custom GPT can do with what you have given it so far. In the Preview pane, enter the following prompt: **What are the specific league names from the SportsWorldCentral fantasy football platform?**

The response you get when you perform this task may be different from what another user would receive. That is because generative AI responses are nondeterministic. (If you make your query too broad, your GPT may hallucinate and provide information that isn't valid.) Figure 15-6 displays an example where the GPT states that it doesn't have specific information about team names, because they are not publicly available. That is correct; to get the team names, it will need to call an API.

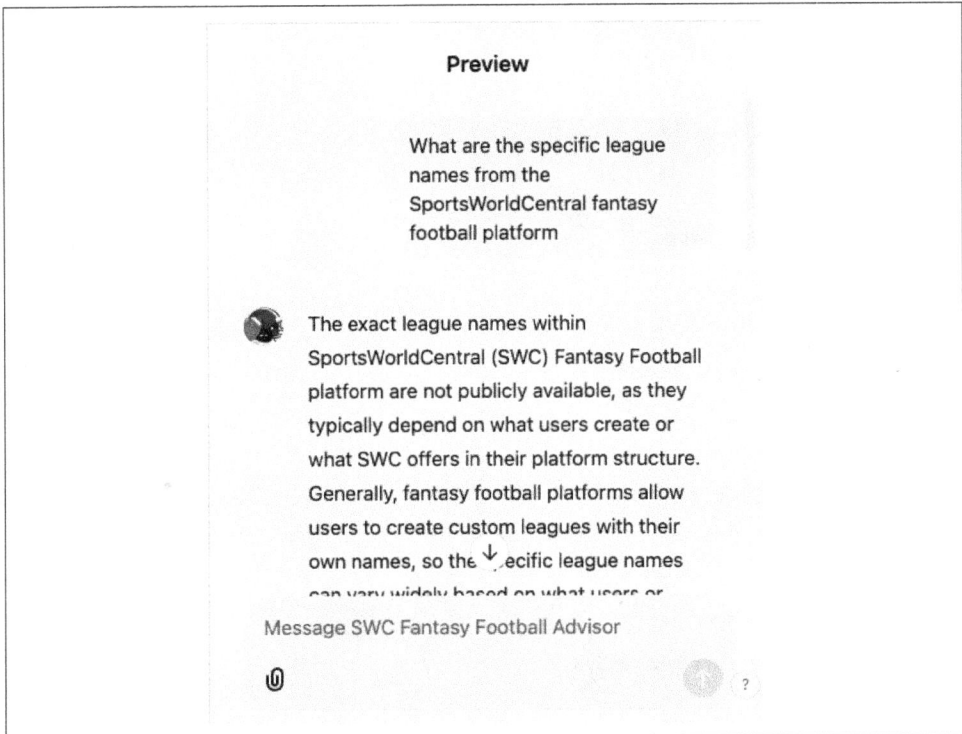

Preview

What are the specific league
names from the
SportsWorldCentral fantasy
football platform

The exact league names within
SportsWorldCentral (SWC) Fantasy Football
platform are not publicly available, as they
typically depend on what users create or
what SWC offers in their platform structure.
Generally, fantasy football platforms allow
users to create custom leagues with their
own names, so the ↓ .ecific league names
can vary widely based on what users or

Message SWC Fantasy Football Advisor

Figure 15-6. Prompt and response from custom GPT

This reflects a *knowledge gap*, which means that the model has not been trained on the information it needs to answer your prompt. To overcome this knowledge gap, you will create a custom action that enables your GPT to call the SWC API.

First, you need to launch your API in GitHub Codespaces with a public endpoint and update it to include the server URL in the OAS file.

Launching Your GitHub Codespace

You will be working with the Part III GitHub Codespace that you created in "Cloning the Part III Repository" on page 240. If you haven't created your Part III Codespace yet, you can complete that section now.

Running the SportsWorldCentral (SWC) API in GitHub Codespaces

In the Terminal of your Codespace, install the required libraries using the provided *requirements.txt* file as follows:

```
.../ai-project (main) $ cd api
.../api (main) $ pip3 install -r requirements.txt
```

Launch the API from the command line as shown:

```
.../api (main) $ fastapi run main.py
```

You will see several messages, ending with the following:

```
INFO:       Started server process [19192]
INFO:       Waiting for application startup.
INFO:       Application startup complete.
INFO:       Uvicorn running on http://0.0.0.0:8000 (Press CTRL+C to quit)
```

You will see a dialog stating "Your application running on port 8000 is available" as shown in Figure 15-7. Click Make Public.

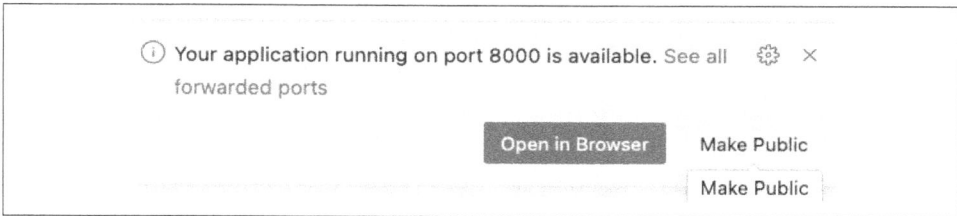

Figure 15-7. Making the API public

The API is now running in Codespaces with a public port. To view the API in the browser, click Ports in the terminal and hover over Port 8000 (see Figure 15-8).

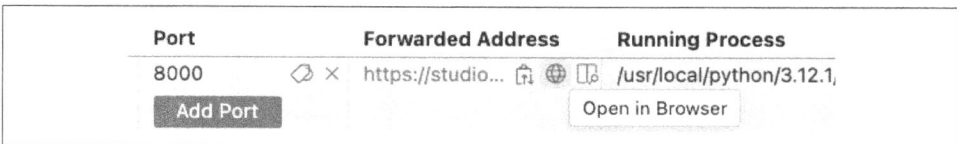

Figure 15-8. API on a public address

Click the globe icon to open in the browser. If you receive the message "You are about to access a development port served by someone's codespace," click Continue.

The browser will open to a URL that ends in *app.github.dev*, and you should see the following health check message in your web browser:

```
{"message":"API health check successful"}
```

Your API is running publicly in the cloud.

Next, you need to capture the public URL of your API so that you can use it in the next section. Copy the URL in the address bar and save that value.

Stop the API by pressing Ctrl-C at the terminal.

Adding the Servers Section to Your OAS File

To update your OAS file, you need to update your FastAPI code. Open the *main.py* file and update the constructor so that it matches the following:

```
app = FastAPI(
    description=api_description,
    title="Sports World Central (SWC) Fantasy Football API",
    version="0.2",
    servers=[
        {"url": "[your base URL from previous step]",
        "description": "Main server"} ❶
    ]

)
```

❶ The value inside the quote marks should be the full public URL you copied in the previous step.

Adding this section in the code makes two changes that you can verify:

- The Servers section will be added to the OAS file.
- The Servers section will be added to the API documentation (which is auto-generated from the OAS file).

To verify these, start your API again from the Terminal:

```
.../api (main) $ fastapi run main.py
```

Navigate to the API documentation by adding */docs* to the URL. Below the General section, you should see a Servers select box as shown in Figure 15-9.

General

Get information about the SWC fantasy football platform as a whole.

Servers

https://didactic-tribble- -8000.app.github.dev - Main server ∨

Figure 15-9. Servers section added to API documentation

To view the OAS file, click the *openapi.json* link at the top of the documentation page. (You may need to add a browser extension to format the JSON for readability.) Below the `info` section of the OAS file, you should see a new `servers` section that looks like the following:

```
"servers": [
    {
        "url": "[your base URL]",
        "description": "Main server"
    }
]
```

Copy the full path of the OAS file from the browser, including *openapi.json*. You will use it in the next step.

Creating a GPT Action

As shown in Figure 15-1, your custom GPT will contain a custom action inside it. This is the component that calls your API.

At the bottom of the Configure tab, click the "Create new action" button. The Add Actions page will be displayed. Leave the Authentication entry as None, because your API does not require authentication.

Click Import URL, and then paste the full address of your OAS file from the previous step. Your display should look like Figure 15-10 at this point.

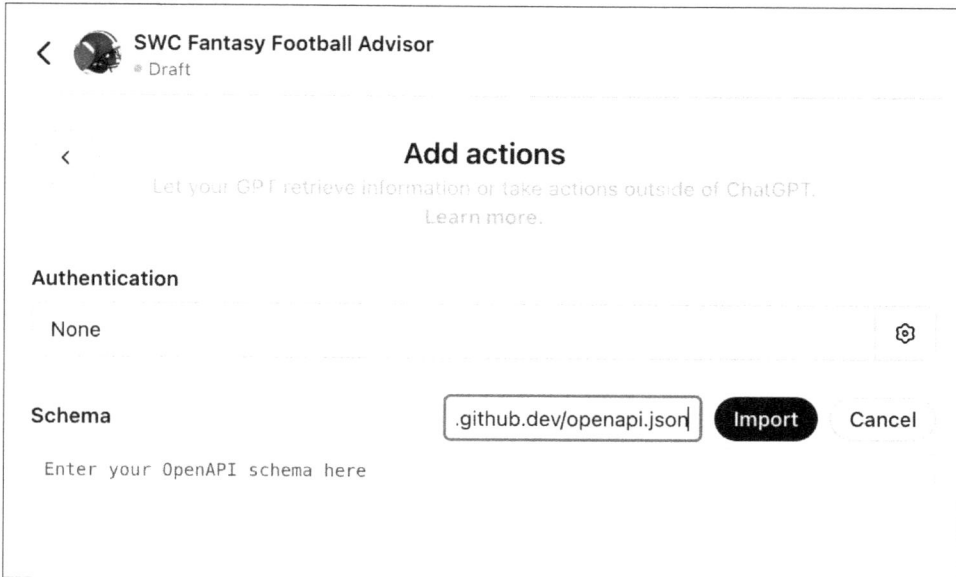

Figure 15-10. Importing the OAS file

Click Import. The Schema section will be populated with the contents of your OAS file, as shown in Figure 15-11.

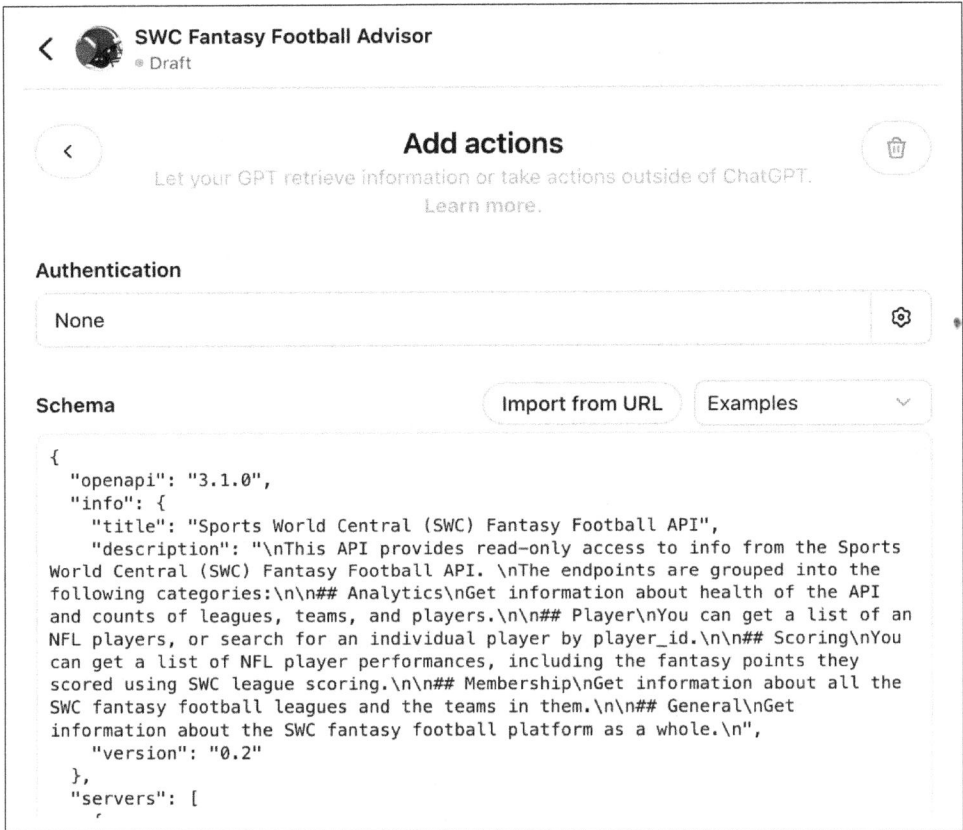

Figure 15-11. OAS file populated

Scroll down below the text of the OAS file and you will see a list of available actions. Figure 15-12 shows the top of this list.

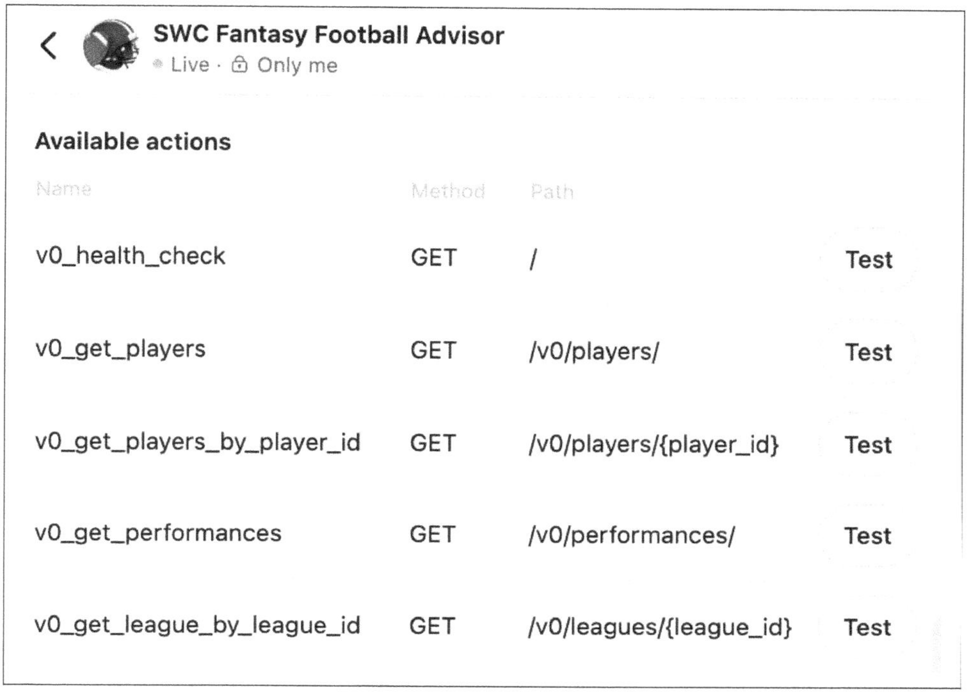

Figure 15-12. Available actions list

Testing the APIs in Your GPT

You will notice that the "Available actions" section lists all of the endpoints in your OAS file. If you wanted to restrict the GPT to a subset of the endpoints in your API, you could manually edit this schema to remove any endpoints you did not want it to use. All of these endpoints are public read-only, so you will leave them in the schema.

Contrasting ChatGPT with LangGraph

This is a good time to contrast the process of providing APIs to a custom GPT in this chapter and "Creating a LangGraph Agent" on page 267. Using LangGraph, you provided tools to the AI application by creating a toolkit with Python Code. Then, you gave that toolkit to a function-calling model to decide which functions to call. The model decided which function to use based on the description in the Python toolkit.

With the custom GPT in this chapter, you provided the OAS file, and the user interface added all of the path entries as possible tools. You have the option of removing any of those items that you did not want the custom GPT to use. Then the custom GPT decided which item to use based on the description in the OAS path object.

To verify that the connection to your API is set up properly, click the Test button next to *v0_health_check*. Your GPT will call the API's health check endpoint and give a status of the API, as shown in Figure 15-13.

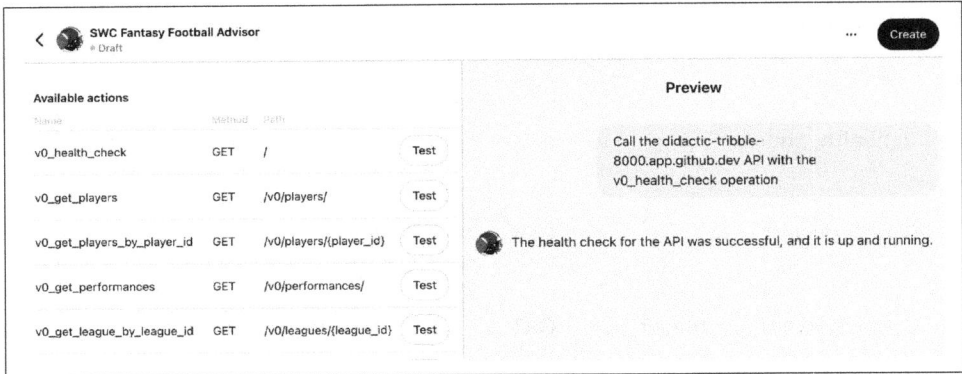

Figure 15-13. GPT calling health check endpoint

You are ready to add this action to your GPT. Click the Save button. The Share GPT dialog will be displayed as shown in Figure 15-14.

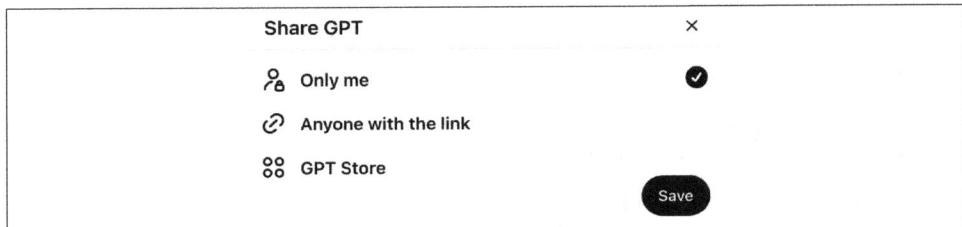

Figure 15-14. Share GPT dialog

Select "Only me" to keep this GPT private. Click Save.

The Settings Saved dialog will be displayed with a direct link to the GPT. You do not need to use the link at this time. Click View GPT.

Chatting with Your Custom GPT

Now that you have created a custom GPT, you will see it displayed in the navigation sidebar beneath ChatGPT. As shown in Figure 15-1, your GPT is a separate application from ChatGPT. Now your GPT has the ability to access the SWC Fantasy Football API to get information.

Select the SWC Fantasy Football Advisor in the navigation. The display should look like Figure 15-15.

SWC Fantasy Football Advisor

By RYAN DAY

This GPT accesses fantasy football APIs to get the latest information.

Message SWC Fantasy Football Advisor

ChatGPT can make mistakes. Check important info.

Figure 15-15. Fantasy Football advisor ready to chat

> Don't miss the note at the bottom of this page: "ChatGPT can make mistakes. Check important info." It bears repeating that conversations with GPTs are impressive and quite convincing, but OpenAI is transparent that LLMs can provide unreliable information and hallucinate in their responses.
>
> Some developers have found that GPTs struggle to process statistics accurately for APIs that return large amounts of data (*https://oreil.ly/Pncx5*). For fantasy sports, the stakes are fairly low. But if your usage expands to tasks where accuracy is more important, don't overlook the warnings that OpenAI is clearly publishing.

Start your conversation by repeating the question you asked before: **What are the specific league names from the SportsWorldCentral Fantasy Football platform?**

The GPT will state that it wants permission to call your API. Click Always Allow.

The GPT will state that it talked to your API. It also will show that it found the `limit` parameter and used a value of 10 to limit the number of results.

Then, it will answer the questions using the information returned from your API, as shown in Figure 15-16. It has overcome the knowledge gap by accessing data from your API.

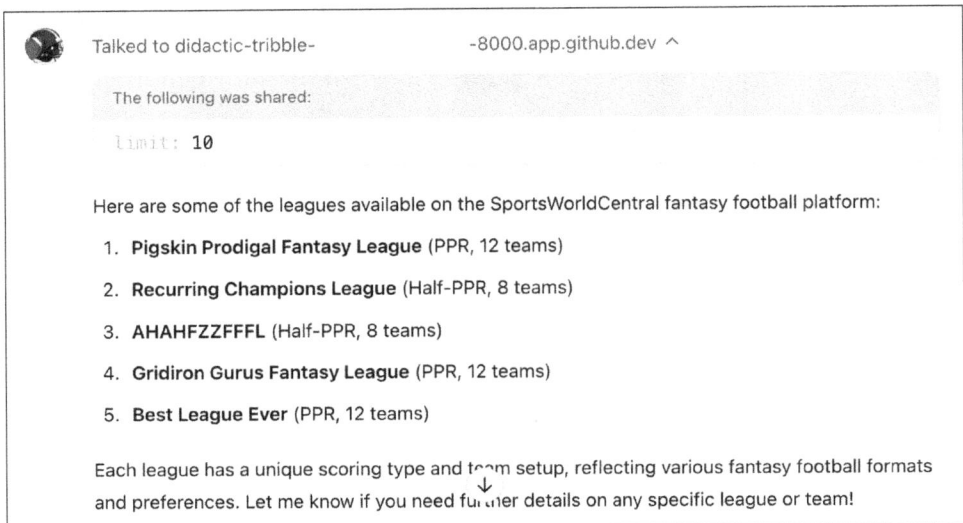

Figure 15-16. Results from API call

To see the API calls that the GPT made, look at the Terminal output in your Codespace. You will see the API calls made:

```
INFO:      10.240.2.131:0 - "GET / HTTP/1.1" 200 OK ❶
INFO:      10.240.4.161:0 - "GET /v0/leagues/?skip=0&limit=100 HTTP/1.1" 200 OK ❷
```

❶ This is a call to the health check endpoint.

❷ This is a call to the get leagues endpoint.

In this case, the custom GPT made the right choices: it used the health check endpoint to make sure the API was available, and it used the get leagues endpoint to see available SWC leagues.

Congratulations! You have created a custom GPT and provided it access to your API through a custom action.

Take time and experiment with questions that will encourage the agent to use more of your API endpoints. Look at the Terminal output to see the API calls that are generated. How can you influence which API it calls? How can you guide it to answer more difficult questions using the tools? How can you stump it?

Completing Your Part III Portfolio Project

You have reached the end of Part III, congratulations! As with Parts I and II, there is some housekeeping required to get the portfolio project ready to share. You'll move the code out of the chapter folders and into functional folders, then update *README.md*.

Before you make these changes, you'll save a copy of your files to a separate GitHub branch, named *learning-branch*, so that it's still available if you want to continue working through the code.

Create the new branch from the command line as follows:

```
.../ai-project/ (main) $ git checkout -b learning-branch ❶
Switched to a new branch 'learning-branch'
.../ai-project/ (main) $ git push -u origin learning-branch ❷
 * [new branch]      learning-branch -> learning-branch
branch 'learning-branch' set up to track 'origin/learning-branch'.
```

❶ Create a new branch named *learning-branch* locally based on the *main* branch.

❷ Push this new branch to your remote repository on GitHub.com.

Next, you will make some changes to the directory structure. Enter the following commands:

```
.../ai-project/ (learning-branch) $ git checkout main ❶
Switched to branch 'main'
Your branch is up to date with 'origin/main'.
.../ai-project/ (main) $ rm -rf chapter13/complete
.../ai-project/ (main) $ rm -rf chapter14/complete
.../ai-project/ (main) $ mkdir model-training ❷
.../ai-project/ (main) $ mv chapter13/* model-training ❸
.../ai-project/ (main) $ mkdir langchain ❹
.../ai-project/ (main) $ mv chapter14/* langchain ❺
.../ai-project/ (main) $ rm -rf chapter13 ❻
.../ai-project/ (main) $ rm -rf chapter14
```

❶ Switch your Codespace back to the *main* branch of your repository.

❷ Make a new directory for the files from Chapter 13.

❸ Move Chapter 13's files to the new folder.

❹ Make a new directory for the files from Chapter 14.

❺ Move Chapter 14's files to the new folder.

❻ Remove all the subdirectories and their files.

To see the directory structure of the completed project, run the following command:

```
.../ai-project (main) $ tree -d  --prune -I 'build|*.egg-info|__pycache__'
.
```

```
├── api
├── langchain
└── model-training
```

```
3 directories
```

Now update the *README.md* file to showcase your work. Here is a start, and then you can add your own thoughts:

```
# AI Portfolio Project
This repository contains program using industry-standard Python frameworks,
based on projects from the book _Hands-on APIs for AI and Data Science_
written by Ryan Day.
```

Now commit these changes to GitHub, and your Part III portfolio project is ready to share with the world. Congratulations on completing your Part III capstone!

Extending Your Portfolio Project

Here are a few ways to continue to expand your knowledge of ChatGPT actions:

- Use the GPT to call APIs with larger and more complicated datasets. Find instances where it fails or makes mistakes, to explore its limitations.
- Use the GPT Builder to make updates to your custom GPT by asking it in advance and telling it to apply the recommended changes directly.
- If you developed an alternative API in Part I of this book, you can follow the steps in this chapter to explore the data as you did for the SWC Fantasy Football API.
- Create a model card or system card for your custom GPT and post in your project's repository.
- Find more ways to demonstrate your learning in this chapter and publish them on your portfolio site, such as capturing a video or writing a blog post.

Summary

In this chapter, you used OpenAI's ChatGPT to interact with your API:

- You created a custom GPT.
- You created a custom action to give your GPT access to the SWC API.
- You updated the contents of your Part III portfolio project.

Index

comma-separated values (CSV), 5
    loading data into tables, 36
    SDK data for bulk download, 148
committing Codespace changes to GitHub,
    25-27
    always stage all changes and commit, 27
    .gitignore file, 25
container image or Docker image, 113
container runtime as Docker, 113
containerization, 173
    Docker, 113
        (see also Docker container deployment)
    Python app containerization article online,
        125
context for LLMs, 239
context manager, 145
Continuous API Management, 2nd Edition
    (Medjaoui, Wilde, Mitra, and Amundsen),
    13
costs for Anthropic Build Plan, 268
costs for ChatGPT, 286
costs per token for AI services, 235
CREATE TABLE (SQL)
    database tables created, 32-34
    primary key, 34
credentials handled safely, 169
    Anthropic API key, 269
    best practices article online, 170
CrewAI, 238
CRISP-DM process, 247-262
    about, 247
    business understanding, 248
    calling API via POST, 260
    data preparation, 248, 251
    data understanding, 248, 249
        player_training_data_full.csv, 249, 250
    deployment, 248, 254-262
        launching API, 259-262
    evaluation of model, 248, 254
    modeling, 248, 251-254
        saving model to ONNX format, 253
        training the model, 253
Crispin, Lisa, 173
crud.py for database access, 45-49
    about, 39
.csv format (see comma-separated values
    (CSV))
Ctrl-C
    to stop API in terminal, 101, 218, 292

    to stop Docker application, 117
cURL, 89
    Apache Airflow installation in Docker, 201
    Swagger UI output, 89
custom metrics, 179
    APIs as data sources, 180
        Joey Greco on, 181
    creating the Shark League Score, 182-195
        about, 182
        API client file created, 185-186
        building the Shark League Score, 195
        calculating League Balance Score, 192
        calculating League Juice Score, 193
        installing new libraries, 184
        Jupyter Notebook code, 188-189
        Jupyter Notebook created, 186
        logging, 188
        running API in Codespace, 176, 184
        software used, 183
        swc_simple_client.py file, 185, 188
        working with your API data, 189-191
    Leeger Python library, 180
cyclical graphs of LangGraph, 268

D

DAGs (directed acyclic graphs), 199
    API health check, 208, 212
    bulk_player_file_load.py, 206
    coding a shared function, 209-211
    creating first DAG, 206-209
    recur-
        ring_player_api_insert_update_dag.py,
        207
    running your DAG, 211
    shared_functions.py file, 209
    tasks, 200
data acquisition
    APIs as data sources
        AI, 233
        custom metrics, 180
    data pipeline sources of data, 198
        (see also data pipelines)
    databases (see database creation and access)
    Kyle Borgognoni on data gathering, 9
    read replica database, 20
    web scrapers, 9, 180
data analyst, 3
data analytics
    about, 3

S

About the Author

Ryan Day is an advanced data scientist at the Conference of State Bank Supervisors (CSBS). He previously led the digital services division for a federal agency, where he advanced cloud computing, data science, and API development initiatives. He is an experienced open source developer who participates in the FastAPI project by performing code reviews and answering questions.

He holds an AWS Solutions Architect certification and is a member of the National Association of Business Economics. He likes to use fantasy football for his data science and coding projects, because it generates reams of interesting data and is a topic that he's been semi-obsessed with for more seasons than he can count.

Colophon

The animal on the cover of *Hands-On APIs for AI and Data Science* is a greater dwarf lemur (*Cheirogaleus major*). Found in the eastern and northern regions of Madagascar, these nocturnal creatures play a vital role in their ecosystems.

Greater dwarf lemurs are smaller than most other lemurs, weighing between 5 and 21 ounces. The average size of their stout bodies is 9.5 inches; their bushy tails are much longer, measuring between 19 to 21.7 inches. Their short, dense fur ranges from gray to warm reddish-brown, and dark rings line their large orb-like eyes, which contain a reflective layer of tissue that helps them see in the dark. They have small, sharp claws on their hands and feet, which allows them to traverse branches.

Greater dwarf lemurs dwell mainly in forests; they sleep during the day in nests that are built out of leaves, twigs, and grass, or in hollowed sections of trees. Their diet consists of fruits, flowers, vine leaves, nectar, and sometimes insects. Because of their diet, these lemurs are considered pollinators and their fruit consumption aids in seed dispersal.

Unfortunately, greater dwarf lemurs are a vulnerable species; deforestation and slash-and-burn agriculture threaten their habitat, and they are often hunted or captured to be kept as pets locally. Habitat restoration and anti-poaching efforts have been made to protect these creatures from harm.

Many of the animals on O'Reilly covers are endangered; all of them are important to the world.

The cover illustration is by José Marzan Jr., based on an antique line engraving from *English Cyclopedia*. The series design is by Edie Freedman, Ellie Volckhausen, and Karen Montgomery. The cover fonts are Gilroy Semibold and Guardian Sans. The text font is Adobe Minion Pro; the heading font is Adobe Myriad Condensed; and the code font is Dalton Maag's Ubuntu Mono.

O'REILLY®

Learn from experts.
Become one yourself.

60,000+ titles | Live events with experts | Role-based courses
Interactive learning | Certification preparation

**Try the O'Reilly learning platform
free for 10 days.**

www.ingramcontent.com/pod-product-compliance
Lightning Source LLC
Chambersburg PA
CBHW080910220326
41598CB00034B/5534